Approaches in Highly Parameterized Inversion: TSPROC, a General Time-Series Processor to Assist in Model Calibration and Result Summarization

By

Stephen M. Westenbroek, U.S. Geological Survey

John Doherty, Flinders University, Adelaide, and Watermark Numerical Computing, Brisbane

John F. Walker, U.S. Geological Survey

Victor A. Kelson, Layne Hydro, Bloomington, Indiana

Randall J. Hunt, U.S. Geological Survey

Timothy B. Cera, St. Johns River Water Management District, Florida

Chapter 7 of
Section C, Computer Programs
Book 7, Automated Data Processing and Computations

Great Lakes Restoration Initiative

Techniques and Methods 7–C7

U.S. Department of the Interior
U.S. Geological Survey

U.S. Department of the Interior
KEN SALAZAR, Secretary

U.S. Geological Survey
Marcia K. McNutt, Director

U.S. Geological Survey, Reston, Virginia: 2012

For more information on the USGS—the Federal source for science about the Earth, its natural and living resources, natural hazards, and the environment, visit http://www.usgs.gov or call 1–888–ASK–USGS.

For an overview of USGS information products, including maps, imagery, and publications, visit http://www.usgs.gov/pubprod

To order this and other USGS information products, visit http://store.usgs.gov

Suggested citation:
Westenbroek, S.M., Doherty, J., Walker, J.F., Kelson, V.A., Hunt, R.J., and Cera, T.B., 2012, Approaches in highly parameterized inversion: TSPROC, a general time-series processor to assist in model calibration and result summarization: U.S. Geological Survey Techniques and Methods, book 7, chap. C7, 79 p., 3 apps.

Preface

Performance of this computer program has been tested and verified for many test cases, although future applications of the program could reveal errors that were not detected in the test cases. Users are requested to notify the U.S. Geological Survey (USGS) if errors are found in the documentation report or in the computer program.

Correspondence regarding the report or program should be sent to:
USGS Wisconsin Water Science Center
8505 Research Way
Middleton, WI 53562–3581
Attention: Stephen M. Westenbroek
Email: *smwesten@usgs.gov*

Although the computer program has been used by the USGS, no warranty, expressed or implied, is made by the USGS or the United States Government as to the accuracy and functionality of the program and related program material, nor shall distribution of this software and source code constitute any such warranty, and no responsibility is assumed by the USGS in connection therewith.

The TSPROC code and other model-related programs are available for downloading from the USGS at the following World Wide Web address: *http://pubs.usgs.gov/tm/tm7c7/.*

Acknowledgements

The code described here was originally authored by John Doherty of Watermark Numerical Computing. The many organizations that supported John during initial development of the TSPROC code are gratefully acknowledged; these organizations include the University of Idaho, U.S. Environmental Protection Agency, the U.S. Geological Survey, the Australian Land and Water Resources Research and Development Corporation, and the Queensland Department of Natural Resources. The authors wish to acknowledge Brian Cade (U.S. Geological Survey) and John Heasley (Resource Analysis Systems) for kindly sharing their C code for calculating hydrologic indices.

Contents

Figures

Tables

Approaches in Highly Parameterized Inversion: TSPROC, a General Time-Series Processor to Assist in Model Calibration and Result Summarization

By Stephen M. Westenbroek, John Doherty, John F. Walker, Victor A. Kelson, Randall J. Hunt, and Timothy B. Cera

Abstract

The TSPROC (Time Series PROCessor) computer software uses a simple scripting language to process and analyze time series. It was developed primarily to assist in the calibration of environmental models. The software is designed to perform calculations on time-series data commonly associated with surface-water models, including calculation of flow volumes, transformation by means of basic arithmetic operations, and generation of seasonal and annual statistics and hydrologic indices. TSPROC can also be used to generate some of the key input files required to perform parameter optimization by means of the PEST (Parameter ESTimation) computer software. Through the use of TSPROC, the objective function for use in the model-calibration process can be focused on specific components of a hydrograph.

Background

This report documents and describes the use of TSPROC (Time Series PROCessor), a software package designed to assist in the calibration of models by editing and distilling time series datasets into more meaningful observations to be used in the optimization objective function. This report is designed to supplement existing literature and documents regarding PEST (Parameter ESTimation) and TSPROC software. A short summary of a selection of existing documents is provided in this section.

TSPROC has existed in one form or another for several years. The Doherty (2008) version of TSPROC included several supplemental tools that are not included or discussed in this report. These additional tools include:

- **adjobs**— adjusts observation weights for different observation groups in a PEST control file by use of a smple user-adjustable expression,

- **plt2smp**—build a site sample file on the basis of a Hydrological Simulation Program—Fortran (HSPF)-generated plot file; used as part of a composite model run by PEST,

- **smpchek**—checks a site sample file (*.ssf) for correctness,

- **smp2hyd**—rewrites the contents of a site sample file for a user-specified list of sites in a form suitable for plotting against time, and

- **smp2vol**— calculates volumes between arbitrary dates and times for flow samples listed in a site sample file.

This report does not document the support software referenced above, because many of the functions are either model-specific or can largely be accomplished by TSPROC software. The application of TSPROC in a model optimization exercise implies the use of PEST software (Doherty 2010a, b). TSPROC is designed to work with two new additions to the PEST family:

- PEST++ (Welter and others, 2012) is a new code that aims to implement the most popular features of the venerable PEST program while shielding the user from some of the trickier implementation details,

- GENIE (Muffles and others, 2012) is a new parallel run manager designed to make it easier to run models on multiple machines enabled by communications over TCP/IP protocol.

The reader should also be aware of publications that give examples of PEST and TSPROC application to surface-water models, including Doherty and Skahill (2006), Skahill and Doherty (2006), Cocca and others (2004), and Doherty and Johnston (2003).

Introduction

TSPROC is a program designed to assist in model parameter estimation, especially surface-water models, by means of the widely used PEST software (Doherty, 2010a, b). TSPROC may also be used to create some of the basic input files required to make use of PEST. In addition to being a utility for working within the PEST environment, TSPROC is a general purpose tool for working with time series and developing statistics and reports from observations and model results.

Surface-water models are capable of simulating daily (or more frequent) discharge values at many locations and for many years. The thousands of data points contained in each hydrograph are serially related, and in the case of surface-water models, more of this type of data is not necessarily better; therefore, comparing statistical indices or summaries between the simulated and observed data is useful. Although parameter estimation may be successfully applied through the use of daily hydrograph data, the resulting model calibration will not necessarily be able to replicate the aspects of the hydrograph pertinent to the project at hand. In this situation, it is necessary to process the data into a form that focuses the calibration process on the aspects of the hydrograph that are most important to the model's intended use. Because thousands or tens of thousands of data records are involved, automation of model post-processing tasks and PEST input file preparation is desirable.

This publication documents the TSPROC software package and provides details about each of the program commands and options in the scripting language. Appendix 1 describes the format of a "site sample file," a generic American Standard Code for Information Exchange (ASCII) file understood by TSPROC that may be used for storing, importing, and exporting time-series data. Appendix 2 describes the development and basic use of TSPROC when packaged as a Python module.

Parameter Estimation and TSPROC

This section discusses general TSPROC capabilities and describes how TSPROC may be used as part of a composite model subject to parameter optimization.

TSPROC Capabilities

TSPROC fills two roles. First, it is a time-series processor, having the ability to perform many different types of operations on observed and model-generated time series. Second, it automates the generation of PEST input files for calibration tasks of arbitrary complexity based on these time series.

Many of the operations performed by TSPROC are designed specifically for use in a model parameter-estimation context. To make valid comparisons between the modeled and observed time series, the model-generated time series must be interpolated to the times when observations were made. Because observations of a particular environmental quantity are often intermittent rather than regular, TSPROC does not assume that any time series that it manipulates has a constant sample interval.

TSPROC makes possible the incorporation of some or all of the following data types into the model-calibration process.

1. **Raw data.** All or part of complete observation time series can be included; TSPROC can facilitate this by interpolating the model-generated series to field observation times.

2. **Processed or filtered data.** Time-series data can be decomposed or filtered using the digital filtering or hydrograph separation techniques. TSPROC includes digital filtering capabilities that allow the separation of high, medium, and low frequency components of any time series. This can be useful in baseflow separation, as implemented in the Doherty (2008) version of TSPROC (Nathan and McMahon, 1990). The version of TSPROC documented in this report also incorporates U.S. Geological Survey (USGS) HYSEP (HYdrograph SEParation) modules that include three different techniques for computing baseflow separation (Sloto and Crouse, 1996). Modeled and observed processed or filtered counterparts can be individually matched through the calibration process.

3. **Accumulated volumes and masses.** Flow volumes and constituent masses can be accumulated between any number of arbitrary dates and times occurring within the model simulation period. Inclusion of volumetric and mass data, calculated on the basis of field observations and on model-generated flows and constituent concentrations (interpolated to field observation times), respectively, can bring numerical stability to the parameter-estimation process and result in more robust estimates of parameter values.

4. **Exceedance-time characteristics.** As with volumetric and mass data, the inclusion of exceedance-time characteristics in the inversion process can decrease the likelihood of numerical instability while it promotes estimation of a realistic set of parameter values. Furthermore, in many modeling applications it is crucial that a model predict exceedance-time characteristics as accurately as possible under future climatic or management conditions.

5. **Summary statistics** and **period statistics.** Basic statistics (mean, sum, median, maximum, minimum, range, and standard deviation) can be calculated from the terms of a time series or functions of these terms over varying time intervals; the period statistics module allows for these statistics to be calculated on a monthly or annual basis over the length of the time series (for example, mean monthly), or as a series composed of monthly values (for example, monthly mean).

6. **Functions of arbitrary complexity** and **data patterns.** New time series may be calculated on the basis of one or more measured or modeled time series. In many instances of model calibration, it may be better to include a comparison of derived time series, rather than raw time series, in the parameter-estimation process. TSPROC allows the user to calculate any number of new time series based on relationships of arbitrary complexity between existing time series. For example, in some calibration contexts, it may be beneficial to compare the log (or some other function) of an observation type with its model-generated counterpart over all or part of the model simulation time. In other contexts, it may be useful to compare a combination of today's and yesterday's flow with the model-generated equivalent of this same quantity. Minimization of the discrepancies between two such composite time series may result in better parameter estimates, as well as better estimates of the uncertainties associated with these parameters, because it incorporates the correlation structure of flow and constituent observations into the parameter-estimation process (Kuczera, 1983). Relationships such as those used by the USGS LOADEST program (Runkel and others, 2004) may be suitable in some cases.

7. **Hydrologic indices.** When TSPROC is applied to surface-water models, hydrologic indices may be used to ensure that the calibration process results in a calibration faithful to particular aspects of the streamflow record. For example, if a model is being built to address the period and severity of low-flow spells under a variety of land-use and climate scenarios, it is important to include indices capturing low-flow characteristics in the model-calibration process.

Model Parameter Estimation Using TSPROC

Although TSPROC can be run as an independent executable program, TSPROC is most useful when run by PEST as part of a composite model. A composite model is made up of two or more executable programs run in succession by means of a batch or script file. When used in this way, TSPROC acts as a model post-processor, carrying out operations of arbitrary complexity on one or more of the time series generated by the model. Similar operations are used with the observation data to develop an objective function when TSPROC is used as a utility to create the PEST control file and instruction file(s) (described in detail in "Generation of Control and Instruction Files"). The processed observations and their model-generated counterparts can then be compared, and the differences between the two (modified by observation weights) are then reduced to a minimum as part of the PEST parameter-estimation process (fig. 1).

Upon startup, PEST first determines the relation between model response and parameter change for all combinations of the two, with the derivatives stored as a Jacobian matrix. During parameter estimation, PEST modifies one or more model parameters on the basis of information gleaned from the Jacobian matrix. Updated model parameters are written to a new model input file by use of the model template file as a map for locating positions within the model input file where the new parameter values are to be substituted. The composite model (a batch or shell script) is then run, which

in turn runs the model code, followed by TSPROC. TSPROC performs the input, processing, and output tasks as specified in the TSPROC control file, generating a composite model output file (TSPROC output file). PEST reads this new output file, with assistance from the instruction file, and calculates a new set of model parameters to try. The process of parameter estimation continues until the difference between the model and observations (objective function) stops decreasing significantly between iterations.

The composite model may contain multiple processing steps, calculated by many different executable codes. A preprocessor, such as the PAR2PAR code, is often added to the composite model to distribute PEST-estimated parameters to many model computational elements (fig. 2). PAR2PAR is part of the PEST package and can also be used to enforce relationships between parameters; for example, the modeler can require that parameter "1" be less than parameter "2," even though they have overlapping absolute ranges.

PAR2PAR may be also be used, for example, if the modeler wishes to distribute a generic parameter estimated by PEST to many hydrologic response units (HRU) in a surface-water model on the basis of the predominant glacial geology of each HRU. PEST writes a PAR2PAR input file on the basis of the user-provided PAR2PAR template file. In turn, PAR2PAR transforms the single generic parameter value into individual values for each HRU, creating a model input file through the use of a model template file.

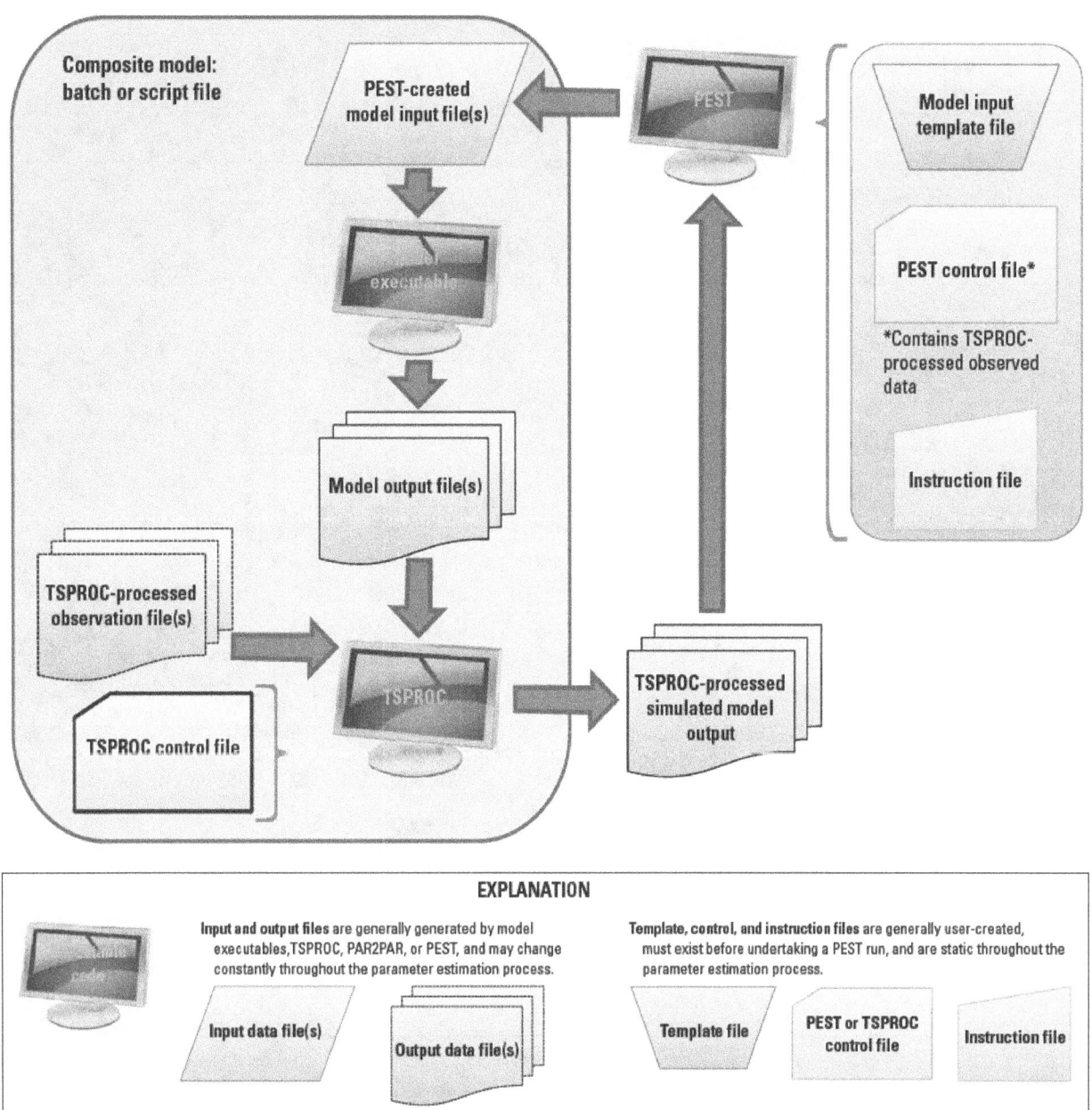

Input and output files are generally generated by model executables, TSPROC, PAR2PAR, or PEST, and may change constantly throughout the parameter estimation process.

Input data file(s)

Output data file(s)

Template, control, and instruction files are generally user-created, must exist before undertaking a PEST run, and are static throughout the parameter estimation process.

Template file

PEST or TSPROC control file

Instruction file

Figure 1. Use of TSPROC as part of a composite model undergoing parameter estimation with PEST.

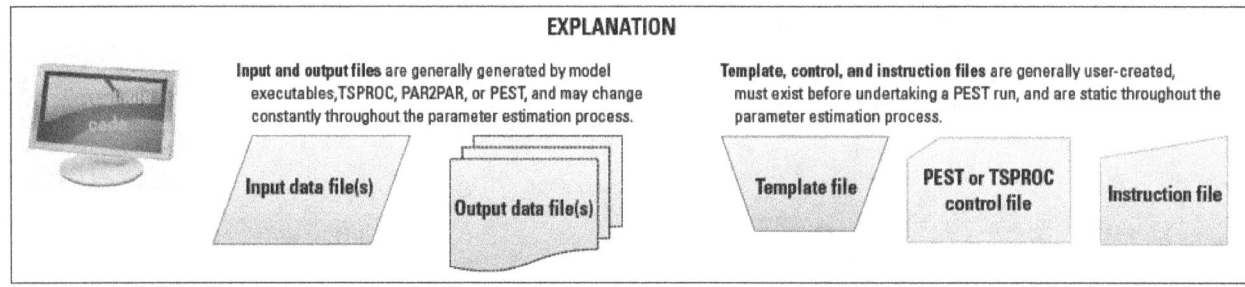

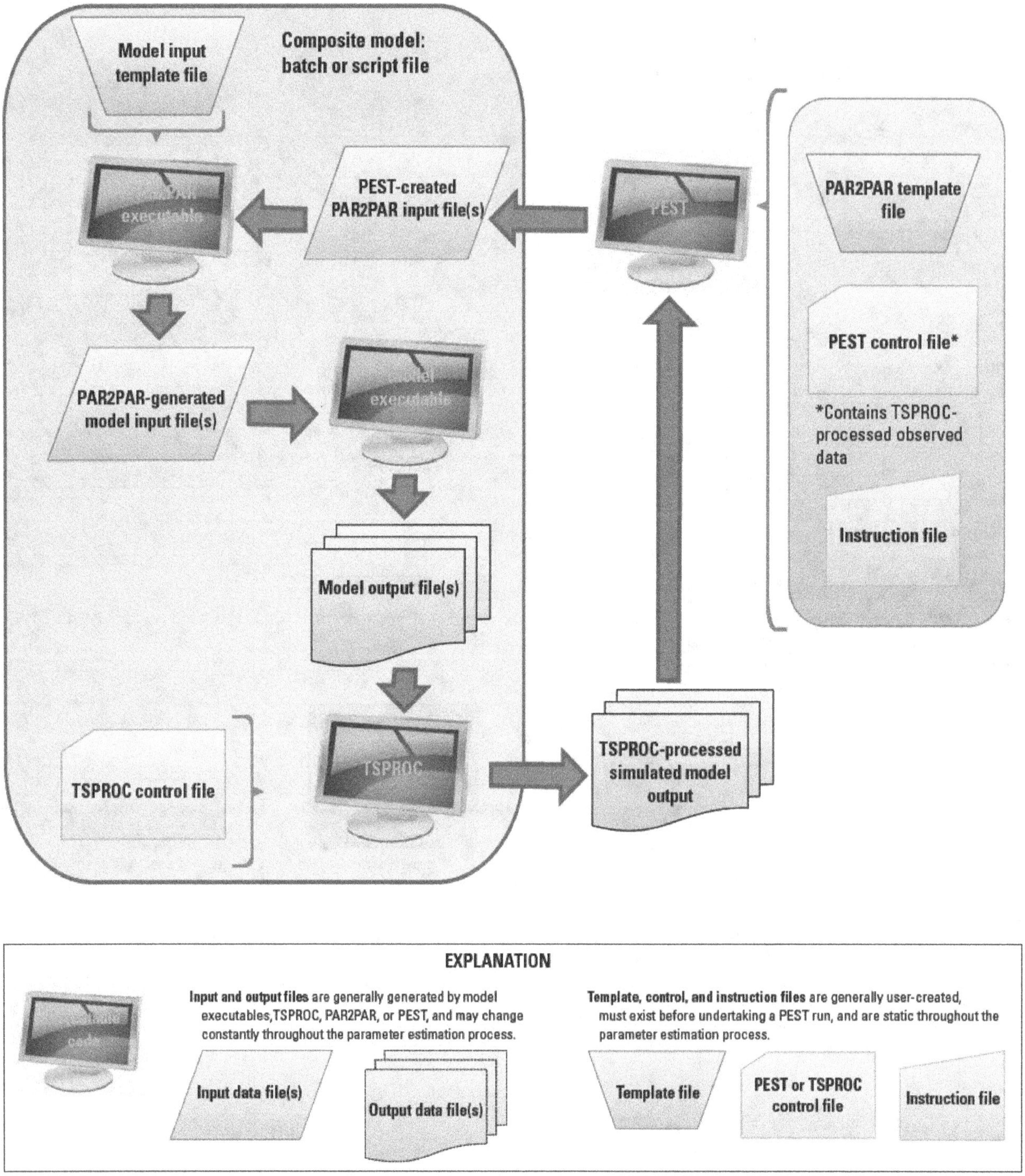

Figure 2.   Use of TSPROC and PAR2PAR as part of a composite model undergoing parameter estimation with PEST.

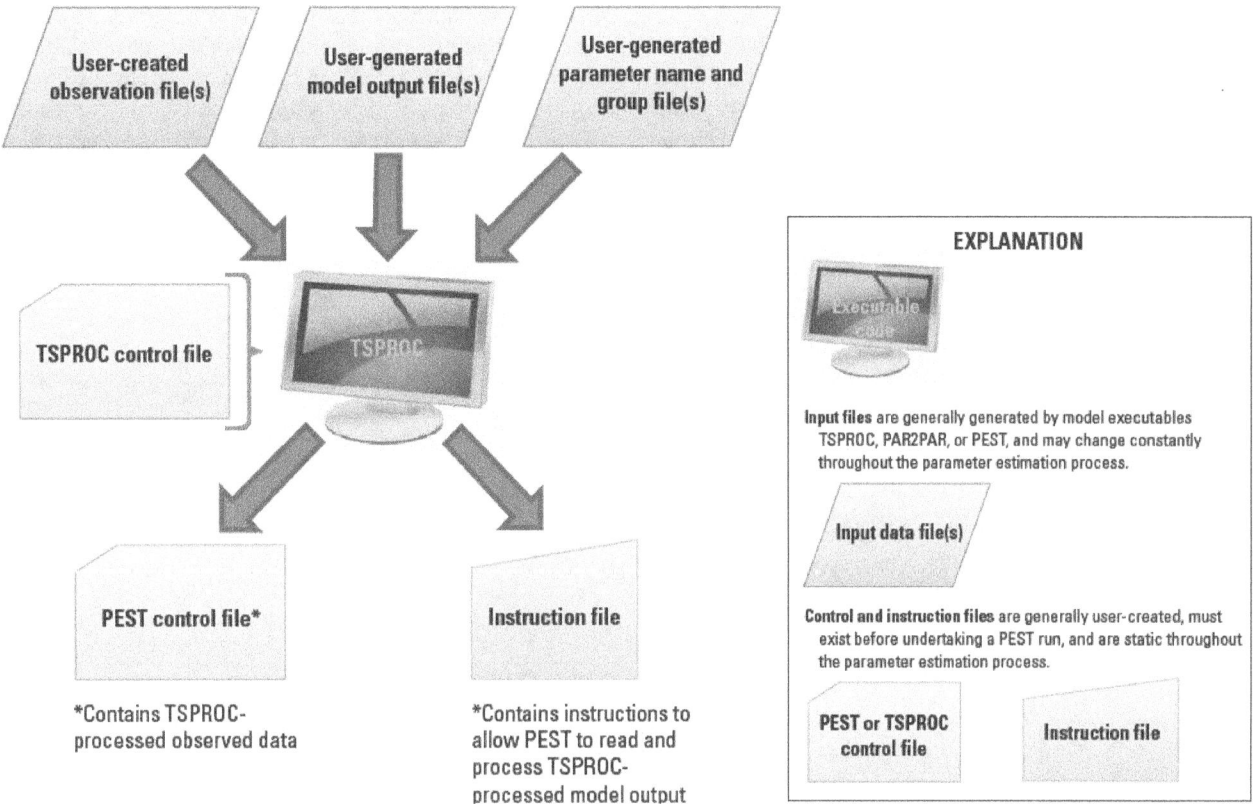

Figure 3. Use of TSPROC to generate a PEST control file and an instruction file.

Generation of Control and Instruction Files

The specific TSPROC block that is run depends on the context. Each block is assigned a context, either an arbitrary name or the special context "all." When TSPROC is run, a context is defined either in the "SETTINGS" block in the TSPROC control file or on the command line. Only those blocks that match the context selected for the run or those blocks with the "all" context are used.

As seen in figures 1 and 2, employing TSPROC as part of the PEST parameter-estimation process can require the use of many template, control, and instruction files. TSPROC can help manage some of this complexity by generating the PEST control file and the instruction file needed to interpret the composite model output (fig. 3).

Raw and processed observation values are embedded in the PEST control file and should not need to be recalculated during model parameter estimation; it thus makes sense to use the TSPROC context mechanism to separate operations involving observed values from those associated with simulated values. A "pest_prep" context might be used to perform input, processing, and output on observed values, with the intermediate results written to an SSF file. An "estimate" context might be used during the actual PEST runs to process only the simulated values, with any required observation series (for example, those needed to provide a time base against which the model simulations are interpolated) read from

the intermediate results SSF file. (See the SETTINGS block section for further discussion of the "context" concept.) If a TSPROC control file contains more than about 10,000 lines of input, much time can be saved by not recalculating statistics on the observed values for each PEST iteration.

Setting up a typical application of TSPROC within a PEST parameter-estimation process includes the following tasks.

1. Processing of various types of observation data to generate an appropriate observation dataset. This task includes reading the observation data from a file, an optional (but recommended) cleaning step to remove questionable or missing data, and as many steps as needed to produce the volumes, statistics, and indices desired in the parameter-estimation process.

2. As part of the composite model, interpolation of raw model-generated time series to the times when field observations were made, then processing of that data in an identical fashion to that used to process the observation data.

3. Generation of PEST input files, including a PEST control file for recording the observations (raw and derived) used in the parameter-estimation process and an instruction file capable of reading the TSPROC-generated output file(s) containing raw and derived simulated values.

To generate PEST input data files, TSPROC must be provided with a set of parameter data and group files pertinent to the current calibration exercise. TSPROC writes a complete PEST control file that records these parameters along with the processed observations to which model-generated equivalents must be matched during the parameter-estimation process. TSPROC assigns names to all observations involved in the parameter-estimation process. Weights are assigned to these observations according to formulas of arbitrary complexity supplied by the user. A different formula can be supplied for each observation type. TSPROC then writes the instruction file informing PEST how to read the composite model-generated equivalents to these processed observations from a TSPROC output file.

Invoking TSPROC

TSPROC is executed from the command line; three optional command line arguments may be supplied. If no arguments are supplied, TSPROC prompts the user for the name of the TSPROC control file and the name of the output run record file. The run record file is a log of what is printed to the screen during the TSPROC run. Figure 4 demonstrates the syntax used when specifying the TSPROC control filename and the run record filename from the command line. Figure 5 demonstrates how the context for the TSPROC run may be overridden when specifying the command and run record filenames.

The TSPROC control file specifies file inputs and outputs as well as the type of time-series processing to be performed on them. As TSPROC runs, it echoes the contents of its input file and the operations that it performs to both the screen and to the run record file. A TSPROC input file is easily prepared by use of a text editor and can have any filename extension allowed by the computer's operating system. The authors have used the extensions "ctl,", "dat,", or "scr" depending on the application and author affiliation.

```
tsproc.exe control_filename run_record_filename
```

Figure 4. Invoking TSPROC—specifying control and run record filenames at the command prompt.

```
tsproc.exe control_filename run_record_filename context_name
```

Figure 5. Invoking TSPROC—specifying control, run record filenames, and context override at the command prompt.

If the context is set to "pest_input", triggering the generation of a set of PEST input files, TSPROC prompt the user before overwriting any existing files of the same name. For example, it may prompt:

```
File instruct.ins already exists. Overwrite it? [y/n]:
```

Type "y" or "n," followed by <Enter> as appropriate. Note that TSPROC does not prompt in this manner when overwriting data files as part of its time-series manipulation functionality; if TSPROC is run many times in the course of a PEST run, these output data files need to be overwritten whenever it is run.

If TSPROC is run by PEST as part of a composite model, it is convenient to supply the control and run record filenames as part of the command syntax used to start a TSPROC run as demonstrated in figure 4.

The Doherty (2008) version of TSPROC did not accept command-line arguments, but instead required placing the responses to TSPROC's prompts into a text file prior to the PEST run and later provided to TSPROC through the command-line redirection mechanism. The version of TSPROC documented in this report may still be invoked in this manner. For example, to invoke TSPROC given an input file named tsproc.ctl and a run record file named tsproc rec, a text file (named, for example, tsproc.in) could be prepared as shown in figure 6.

```
tsproc.ctl
tsproc.rec
```

Figure 6. Contents of a text file containing the responses to TSPROC prompts.

When TSPROC is then run as part of a composite model by PEST, the composite model batch file should invoke TSPROC in one of the ways shown in figure 7:

```
::  The suggested syntax for starting TSPROC from with a batch
::  file is shown in the following line:
tsproc.exe tsproc.ctl tsproc.rec

::  The suggested syntax for starting TSPROC from with a batch
::  file with specified context 'pest_prep' is shown in the
::  following line:
tsproc.exe pest_prep tsproc.ctl tsproc.rec

::  The traditional way to start TSPROC from within a batch file is:
tsproc.exe < tsproc.in
::  Note—you cannot override context using the traditional method
::  The file "tsproc.in" would contain two lines of information
::  specifying the name of the control file and run record file:
tsproc.ctl
tsproc.rec
```

Figure 7.   Example batch file snippets for starting a TSPROC run.

In the third example, the "<" symbol instructs TSPROC to look for its keyboard input from the ASCII file whose name follows it. Of course, the above command can be issued from the keyboard as well, if file tsproc.in has already been prepared.

Any errors encountered while running TSPROC causes a report to be written to the screen and the run record file. Once an error is detected, TSPROC does not read any more lines from the the input file nor does it perform any operations beyond that where the error occurred. Thus, although TSPROC's error-checking functionality is quite comprehensive, it finds only one error at a time in a TSPROC input file that contains multiple errors.

In some circumstances, a user may not want TSPROC to report its activities to the screen, for example if TSPROC is being run under the control of PEST and the user does not want TSPROC screen output to interfere with that of PEST. As with any command-line program, TSPROC output can be re-directed from the screen to a file, leaving the screen bare. Because the TSPROC run record file contains all of the information that TSPROC writes to the screen, nothing is gained through keeping a file that contains redirected screen output. The TSPROC screen output can instead be redirected to the "nul" file ("/dev/null" on Linux®), a special file that accepts all input and does nothing. If is the user desires that TSPROC look to a file named tsproc.in for its keyboard input, and that it redirect its screen output to the "nul" file, it should be run using the syntax as shown in figure 8.

```
tsproc.exe < tsproc.in > nul
```

Figure 8.   Command-line redirection prevents screen output from being written.

Using TSPROC with Unsupported Models

If TSPROC is to be used in the calibration of a model for which it is presently incapable of directly importing results, a workaround solution is to write a small translation program that converts the outputs of the unsupported model to site sample file format. This translation program would be run between the model and TSPROC as part of a composite model calibrated by PEST.

Overview of the TSPROC Input File

The TSPROC input file is divided into a series of sections or "blocks." Within each block, various items of information are supplied following pertinent "keywords" that identify each such item. In most blocks, these keywords can be supplied in any order, although some exceptions to this rule are highlighted elsewhere in this report. TSPROC gives an appropriate error message if keyword ordering is incorrect. Although keywords are capitalized in the illustrations used throughout this report for ease of recognition, the contents of a TSPROC input file are case-insensitive, except for filenames on Linux®.

Any line within a TSPROC input file beginning with the "#" character is ignored, enabling free mixing of comments with data elements in a TSPROC input file. The complex nature of the instructions that can be supplied to TSPROC through its input file makes the inclusion of comments in this file a good idea. Figure 9 gives a short example showing the contents of a TSPROC control file.

Each block within a TSPROC input file instructs TSPROC to carry out a certain type of operation. Information supplied within a block informs TSPROC of the names of the entities to be processed and the names of the entities to be produced as a result of that processing. Any other information required by TSPROC to enable that processing to take place is also supplied within the block through the appropriate keyword. For each block, some keywords are optional and some are mandatory. Where an optional keyword is not supplied, TSPROC supplies a default value for its associated variable.

With the exception of the "SETTINGS" block, blocks may be arranged in a TSPROC input file in any order. However, because TSPROC processes blocks in the order in which they are supplied, the ordering of blocks can be important in many applications. If a sequence of operations is carried out, they must occur in the order in which they must be processed; for example, a reading step must occur before a cleaning step, which must occur before statistics are calculated.

```
##########################################################################
# The 'SETTINGS' block must be first within the control file.
##########################################################################

START SETTINGS
 CONTEXT pest_input
 DATE_FORMAT mm/dd/yyyy
END SETTINGS


##########################################################################
# Modeled river flows are read from a HSPF output file.
##########################################################################

START GET_MUL_SERIES_PLOTGEN
 CONTEXT all
 FILE catchment.plt
 LABEL "total outflow"
 NEW_SERIES_NAME flow_mod
END GET_MUL_SERIES_PLOTGEN


##########################################################################
# Observed river flows are read from a WDM file.
##########################################################################

START GET_SERIES_WDM
 CONTEXT all
 FILE catchment.wdm
 DSN 113
 NEW_SERIES_NAME flow_obs
END GET_SERIES_WDM


##########################################################################
# Modeled flows are interpolated to the times of observed flows.
##########################################################################

START NEW_TIME_BASE
 CONTEXT all
 SERIES_NAME flow_mod
 TB_SERIES_NAME flow_obs
 NEW_SERIES_NAME i_flow_mod
END NEW_TIME_BASE


##########################################################################
# Flow volumes are accumulated for the modeled time series.
##########################################################################

START VOLUME_CALCULATION
 CONTEXT all
 SERIES_NAME i_flow_mod
 NEW_V_TABLE_NAME vol_mod
 FLOW_TIME_UNITS days
 DATE_FILE dates.dat
END VOLUME_CALCULATION
```

Figure 9. Example of a TSPROC control file.

```
####################################################################
# Flow volumes are accumulated for the observed time series.
####################################################################

START VOLUME CALCULATION
 CONTEXT pest input
 SERIES NAME flow obs
 NEW V TABLE NAME vol obs
 FLOW TIME UNITS days
 DATE FILE dates.dat
END VOLUME CALCULATION


####################################################################
# Exceedance times are calculated for the modeled time series.
####################################################################

START EXCEEDANCE TIME
 CONTEXT all
 SERIES NAME i flow mod
 NEW_E_TABLE_NAME time_mod
 EXCEEDANCE TIME UNITS days
 FLOW 0
 FLOW 10
 FLOW 20
 FLOW 50
 FLOW 100
 FLOW 200
END EXCEEDANCE TIME


####################################################################
# Exceedance times are calculated for the observed time series
####################################################################

 START EXCEEDANCE TIME
 CONTEXT pest input
 SERIES NAME flow obs
 NEW E TABLE NAME time obs
 EXCEEDANCE_TIME_UNITS days
 FLOW 0
 FLOW 10
 FLOW 20
 FLOW 50
 FLOW 100
 FLOW 200
END EXCEEDANCE TIME
```

Figure 9. Example of a TSPROC control file.—Continued

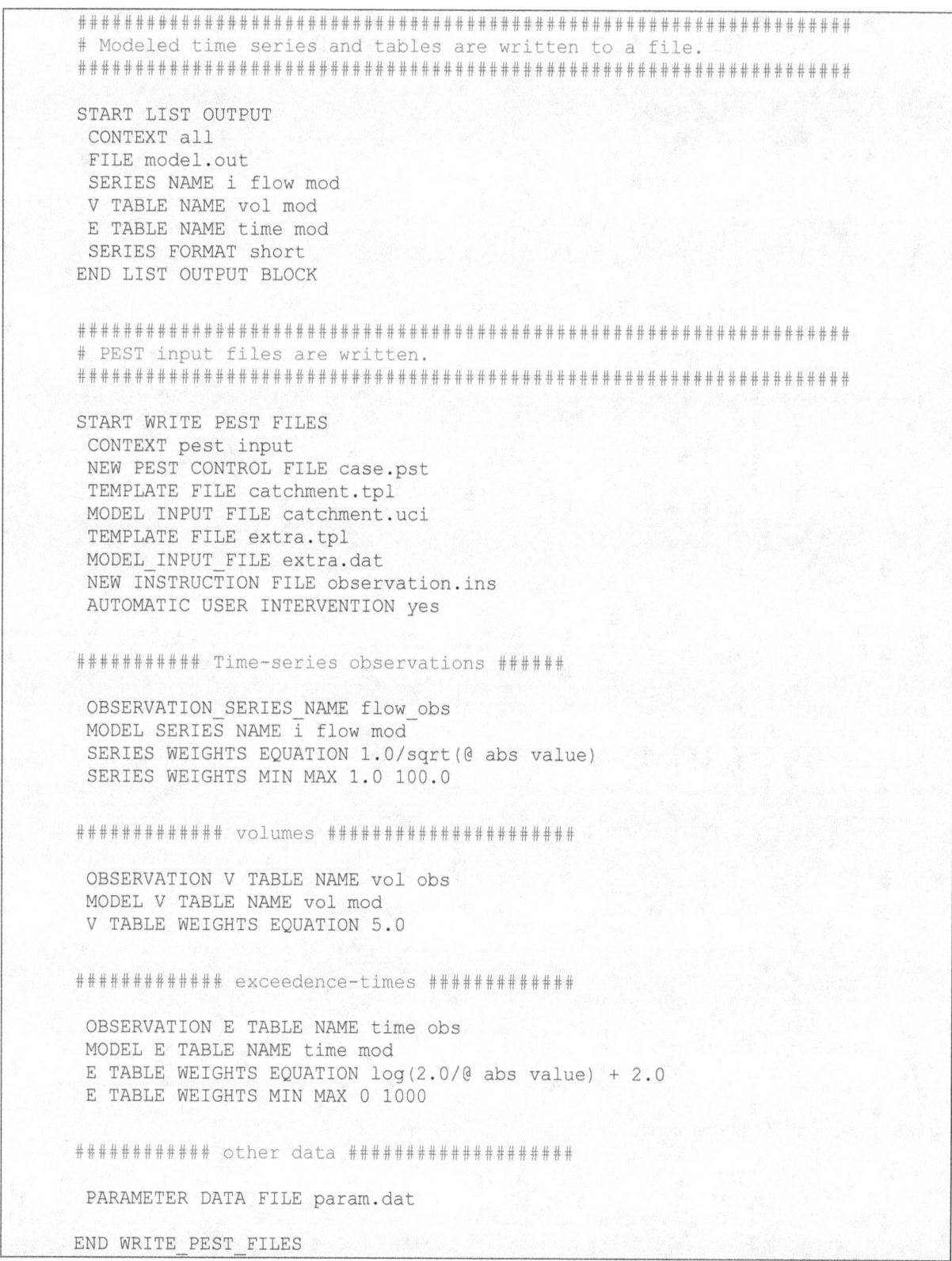

```
##########################################################################
# Modeled time series and tables are written to a file.
##########################################################################

  START LIST OUTPUT
   CONTEXT all
   FILE model.out
   SERIES NAME i flow mod
   V TABLE NAME vol mod
   E TABLE NAME time mod
   SERIES FORMAT short
  END LIST OUTPUT BLOCK


##########################################################################
# PEST input files are written.
##########################################################################

  START WRITE PEST FILES
   CONTEXT pest input
   NEW PEST CONTROL FILE case.pst
   TEMPLATE FILE catchment.tpl
   MODEL INPUT FILE catchment.uci
   TEMPLATE FILE extra.tpl
   MODEL_INPUT_FILE extra.dat
   NEW INSTRUCTION FILE observation.ins
   AUTOMATIC USER INTERVENTION yes


########### Time-series observations ######

   OBSERVATION_SERIES_NAME flow_obs
   MODEL SERIES NAME i flow mod
   SERIES WEIGHTS EQUATION 1.0/sqrt(@ abs value)
   SERIES WEIGHTS MIN MAX 1.0 100.0


############# volumes #####################

   OBSERVATION V TABLE NAME vol obs
   MODEL V TABLE NAME vol mod
   V TABLE WEIGHTS EQUATION 5.0


############# exceedence-times #############

   OBSERVATION E TABLE NAME time obs
   MODEL E TABLE NAME time mod
   E TABLE WEIGHTS EQUATION log(2.0/@ abs value) + 2.0
   E TABLE WEIGHTS MIN MAX 0 1000


############ other data ###################

   PARAMETER DATA FILE param.dat

  END WRITE_PEST_FILES
```

Figure 9.   Example of a TSPROC control file.—Continued

The SETTINGS Block

In any TSPROC input file, the SETTINGS block must come before all other blocks. The SETTINGS block must contain two keywords: DATE_FORMAT and CONTEXT.

The DATE_FORMAT keyword informs TSPROC of the protocol to be used for representation of dates in all input files that it reads and output files that it generates. In the abbreviated date formats entered into the block, two digits represent the date, two digits represent the month, four digits represent the year, and a slash character separates each set of digits. Only two options are presently available: dd/mm/yyyy and mm/dd/yyyy.

The CONTEXT keyword must be followed by a character string of 20 characters or less (with no embedded spaces) that sets the "context" for the current TSPROC run. A CONTEXT keyword is also a mandatory element of each block appearing in a TSPROC input file; as in the SETTINGS block, the CONTEXT keyword in all of these blocks must be followed by a string of 20 characters or less. Up to five CONTEXT keywords can appear in any TSPROC processing block. (A "processing block" is any block other than the SETTINGS block.) If the CONTEXT string following any of the CONTEXT keywords in a processing block agrees with that in the SETTINGS block, then the instructions in that block are processed by TSPROC. If not, the block is ignored.

If at least one of the CONTEXT strings supplied in a processing block is "all," the operations listed in that block override the current TSPROC context defined in the SETTINGS block. Furthermore, CONTEXT keywords must precede all other keywords in the processing block. Use of the CONTEXT concept allows a user to "turn on" and "turn off" various processes cited in a TSPROC input file simply by altering the CONTEXT string in the SETTINGS block. Thus, the same TSPROC input file can function in both the preparation for and the execution of the calibration process, which can be very useful when preparing for a PEST run.

Specifying TIME and DATE within a Block

Many of the processing blocks in TSPROC allow for the optional processing of a subset of the available values based on a user-specified time and date range. User-specified values for DATE_1 and TIME_1 define the starting point in the range, and values specified for DATE_2 and TIME_2 define the endpoint of the range. If the DATE and TIME arguments are omitted, then the entire time series is processed. If a DATE_1 keyword is present but a TIME_1 keyword is absent, then TIME_1 is assumed to be 00:00:00; the same holds for DATE_2. If TIME_1 is present, then DATE_1 must be present; the same holds for TIME_2.

Blocks within a TSPROC Input File

Table 1 lists the blocks that may be present within any TSPROC input file. Multiple occurrences of any block except the SETTINGS block are permitted.

Each time series must be given a name by the user when it is imported into TSPROC or produced as an outcome of the processing encapsulated in a TSPROC processing block; a time series is normally named with a NEW_SERIES_NAME keyword. A time series name must be 18 or fewer characters.

Many of the TSPROC processing options produce a new time series through processing or manipulating one or a number of existing time series. Where this occurs, the user must provide the name of both the existing time series (through a SERIES_NAME keyword) and the new time series (through a NEW_SERIES_NAME keyword) to the current processing block.

Sometimes the processing of a time series results in the creation of an entity that is not another time series. When TSPROC calculates certain statistics pertaining to the terms of a time series (through the SERIES_STATISTICS block), these statistics are stored in an "S_TABLE." The outcomes of volumetric calculations carried out by the VOLUME_CALCULATION block are stored in an "V_TABLE.table," The outcomes of exceedance time calculations carried out by the EXCEEDANCE-TIME block are stored in an "E_TABLE.table." Statistics based on the comparison of two time series are written to a "C_TABLE." Hydrologic indices are stored in a general table (G_TABLE). As for the time-series entity, each of these other entities must be assigned a name of 18 or fewer characters that follows the NEW_C_TABLE, NEW_S_TABLE, NEW_V_TABLE, NEW_G_TABLE and NEW_E_TABLE keywords in the pertinent processing blocks

TSPROC never overwrites one entity with another; the name provided for a new entity in a processing block must be different from the name of any existing entity of the same type. If desired, the ERASE_ENTITY block can be used to erase entities from memory to make room for other entities. This functionality can be very important when processing many lengthy time series that make large demands on computer memory.

The remainder of this report section contains detailed discussions of the TSPROC blocks, arranged in alphabetical order. Each block must contain one to five CONTEXT keywords and arguments as the first entries in the block; other keyword entries may appear in any order within a block, unless noted otherwise.

Table 1. TSPROC input blocks.

[HSPF, Hydrological Simulation Program-Fortran; USGS, U.S. Geological Survey; GSFLOW, Coupled Groundwater-Surface Water Flow Model; PRMS, Precipitation-Runoff Modeling System; WDM, Watershed Data Management; UFORE_HYDRO, Urban Forest Effects Hydrological model; PEST, Parameter ESTimation software]

Block name	Function of block
Required	
SETTINGS	Provides settings for the current TSPROC run.
Input	
GET_MUL_SERIES_PLOTGEN	Imports one or more time series from a HSPF PLOTGEN file.
GET_MUL_SERIES_GSFLOW_GAGE	Imports one or more series from "gage files" produced by the USGS GSFLOW model.
GET_MUL_SERIES_SSF	Imports multiple time series from a site sample file.
GET_MUL_SERIES_STATVAR	Imports one or more series from "statvar files" written by the PRMS model.
GET_SERIES_SSF	Imports a time series from a site sample file.
GET_SERIES_TETRAD	Imports one or more time series from a TETRAD output file.
GET_SERIES_UFORE_HYDRO	Imports a time series from a UFORE_HYDRO model input or output file.
GET_SERIES_WDM	Imports a time series from a WDM file.
General series and table management	
ERASE_ENTITY	Removes a time series or table from TSPROC memory.
NEW_SERIES_UNIFORM	Creates a uniform valued series with a constant time increment.
NEW_TIME_BASE	Interpolates one time series to the sample dates and times of another.
REDUCE_TIME_SPAN	Shortens a time series by deleting terms outside a user-specified date/time interval.
SERIES_BASE_LEVEL	Subtracts a single term of one time series from all terms of another time series.
SERIES_CLEAN	Erases terms in a series between user-supplied thresholds.
SERIES_COMPARE	Calculates statistics that describe the goodness of fit between one time series and another, placing the results in a C_TABLE.
SERIES_DIFFERENCE	Computes a new time series as the difference of subsequent terms of an existing time series.
SERIES_DISPLACE	Advances or retards the terms of a time series by a multiple of the sample interval.
V_TABLE_TO_SERIES	Copies data from a V_TABLE to a time series.
Statistics and specialized processing	
DIGITAL_FILTER	Passes a time series through a high pass, low pass, or band pass digital Butterworth filter, or a "base flow separation filter," to produce a new time series.
EXCEEDANCE_TIME	Calculates the times beyond which terms of a time series exceed user-specified thresholds, creating an E_TABLE.
HYDRO_EVENTS	Extracts portions of a time series associated with peak flow events.
HYDROLOGIC_INDICES	Computes hydrologic indices for a time series.
PERIOD_STATISTICS	Calculates basic statistics on a time series on a monthly or annual basis.
SERIES_EQUATION	Carries out mathematical operations of arbitrary complexity between the terms of any number of time series of identical time base to create a new time series.
SERIES_STATISTICS	Calculates certain statistics based on some or all of the terms comprising a time series, creating an S_TABLE.
USGS_HYSEP	Performs hydrograph separation on the time series using one of three published USGS methods.
VOLUME_CALCULATION	Calculates volume or mass by time integration of a flow or flux time series, thus creating a V_TABLE.
Output	
LIST_OUTPUT	Writes TSPROC time series and tables to a text or site sample file.
WRITE_PEST_FILES	Generates a PEST control file and a PEST instruction file for a parameter estimation process that includes any number of time series, S_TABLEs, V_TABLEs and E_TABLEs.

DIGITAL_FILTER

The DIGITAL_FILTER block instructs TSPROC to calculate a new time series by passing an existing series through a digital filter. Two types of filter are provided. The Butterworth filter can remove high frequency components (low pass filter), low frequency components (high pass filter), or both (band pass filter) from the original time series. The baseflow separation filter allows extraction of quickflow (or storm flow) from a flow time series; baseflow can then be obtained by subtracting quickflow from the original series by means of a SERIES_EQUATION block.

Digital filtering can only be performed on a time series that has a constant sample interval and enough terms for the filtering algorithm to work. Thus, before performing filtering operations, TSPROC checks the specified time series for these conditions; if they are not met, TSPROC terminates execution with an error message. (Use the NEW_TIME_BASE block with the NEW_SERIES_UNIFORM block to create a time series interpolated to a uniform time base, if this is a problem.) TSPROC does not allow filtering operations to take place on a time series that has fewer than 20 data values.

Keywords available in the DIGITAL_FILTER block are listed in table 2. Example application of a Butterworth filter within a DIGITAL_FILTER block is shown in figure 10; application of a baseflow separation filter (Nathan and McMahon, 1990) is shown in figure 11.

Table 2. Keywords in a DIGITAL_FILTER block.—Continued

[db, decibel]

Keyword	Role	Specifications
CONTEXT	At least one CONTEXT keyword must be supplied; up to five are permitted. If one of the CONTEXT strings matches the CONTEXT string in the SETTINGS block, or if one of the CONTEXT strings is "all," the DIGITAL_FILTER block will be processed.	Any character string without internal spaces of 20 characters or less in length. The CONTEXT keyword(s) must precede all other keywords.
SERIES_NAME	Mandatory. The name of the time series on which filtering operations will be carried out.	A name of 18 or fewer characters referencing a time series stored within TSPROC's memory.
NEW_SERIES_NAME	Mandatory. The name of the new series created by TSPROC through filtering of an existing time series.	Any character string without spaces up to 18 characters.
FILTER_TYPE	Mandatory. The type of filter being implemented.	"butterworth" or "baseflow_separation" .
FILTER_PASS	Mandatory if FILTER_TYPE is "butterworth"; disallowed otherwise. Informs TSPROC whether to carry out low, band, or high pass filtering.	"low," "band," or "high".
CUTOFF_FREQUENCY	Mandatory if FILTER_TYPE is "butterworth" and FILTER_PASS is "high" or "low"; disallowed otherwise. For a high pass filter, the 3dB point of low frequency roll-off. For a low pass, filter the 3dB point of high frequency roll-off. Frequency in days: 1.	Real number.
CUTOFF_FREQUENCY_1	Mandatory if FILTER_TYPE is "butterworth" and FILTER_PASS is "band"; disallowed otherwise. The 3dB point of low frequency roll-off. Frequency in days 1.	Real number.
CUTOFF_FREQUENCY_2	Mandatory if FILTER_TYPE is "butterworth" and FILTER_PASS is "band"; disallowed otherwise. The 3dB point of high frequency roll-off. Frequency in days 1.	Real number.
STAGES	Optional if FILTER_TYPE is "butterworth"; disallowed otherwise. Number of filter stages. The more stages, the steeper is the high and/or low frequency roll-off. Default is 1.	Integer. 1, 2, or 3.
ALPHA	Mandatory if FILTER_TYPE is "baseflow_separation"; disallowed otherwise. The assumed relative decay rate of baseflow.	A real number greater than zero (normally in the range 0.9 to 0.975).

Table 2. Keywords in a DIGITAL_FILTER block.—Continued

[db, decibel]

Keyword	Role	Specifications
PASSES	Optional if FILTER_TYPE is "baseflow_separation"; disallowed otherwise. The number of filter passes. Default is 1.	Integer. 1 or 3 only.
REVERSE_SECOND_STAGE	Optional. If FILTER_TYPE is set to "butterworth," STAGES is set to 2, and FILTER_PASS is set to "low," then the second filter pass is performed in the reverse direction, thereby nullifying any phase shift incurred in the first low pass filter pass.	"yes" or "no".
CLIP_INPUT	Optional for baseflow separation filter type; disallowed for butterworth. If activated, prevents terms of filtered time series from exceeding terms of original time series. Default is "no."	"yes" or "no".
CLIP_ZERO	Optional for baseflow separation filter type; disallowed for butterworth. If activated, prevents terms of filtered time series from becoming negative. Default is "no."	"yes" or "no".

```
    START DIGITAL FILTER
    CONTEXT context_1
    FILTER_TYPE butterworth
    SERIES_NAME flow
    NEW_SERIES_NAME av_flow
    FILTER PASS low
    CUTOFF_FREQUENCY 0.08
    END DIGITAL_FILTER
```

Figure 10. Example of a DIGITAL_FILTER block for applying a Butterworth filter.

```
    START DIGITAL FILTER
    CONTEXT context_1
    FILTER_TYPE baseflow_separation
    SERIES_NAME flow
    NEW_SERIES_NAME qflow
    ALPHA 0.95
    PASSES 1
    CLIP_INPUT yes
    CLIP_ZERO yes
    END DIGITAL_FILTER
```

Figure 11. Example of a DIGITAL_FILTER block for applying a baseflow_separation filter.

Digital filtering is a fast and powerful means of accentuating certain aspects of a time series and removing others. A high pass filter removes long-term variations from a time series, while a low pass filter removes short-term variations. A band pass filter removes both short- and long-term variations, allowing only medium-term variations to remain in the filtered time series. Many different types of filters can be constructed to implement all three of these types of operation. TSPROC implements the Butterworth filter; this has the desirable property that its frequency response is maximally flat within the pass band. TSPROC also implements a baseflow separation filter— a form of high pass filter with a more gentle frequency roll-off than the Butterworth filter outside the pass band. The baseflow separation filter is suitable for separation of the quickflow component of streamflow; baseflow can then be obtained by subtraction from the original streamflow. See Nathan and McMahon (1990) for details.

Use of each of the types of digital filter implemented by TSPROC is discussed in the following sections.

Butterworth Filter

When a Butterworth filter is used, its frequency characteristics must be provided directly through pertinent keywords within the DIGITAL_FILTER block.

The boundary between the pass band and the stop band of a filter is normally denoted by the "3 decibel point," because 3 decibels (dB) is the frequency at which the amplitude response is a factor of about $\sqrt{2}$ less than it is in the pass band. In designing a low pass Butterworth filter, one such frequency is required, and it is supplied with the CUTOFF_FREQUENCY keyword. The same holds for a high pass Butterworth filter, except that the amplitude decreases with decreasing frequency from the 3dB point for a high pass filter, whereas it decreases with increasing frequency from the 3dB point for a low pass filter. For a band pass filter, an upper and a lower 3dB frequency are required and must be supplied following the CUTOFF_FREQUENCY_1 and CUTOFF_FREQUENCY_2 keywords. The former must be less than the latter, or TSPROC terminates processing of the DIGITAL_FILTER block and issues an error message.

Frequencies must be supplied in units of day^{-1} regardless of the time increment of the time series. Sometimes it is easier to think in terms of period rather than frequency; period is the reciprocal of frequency. A fluctuation that repeats itself every n days has a frequency of $1/n$ day^{-1}; note than n can be greater or less than a day. For a period of 6 hours, n is ¼ days and the frequency is 4 day^{-1}; for a period of 10 days, the frequency is $1/10$ day^{-1}.

To prevent aliasing (the presence of unwanted components in the filtered signal), a high, low, or band pass cutoff frequency must be less than one-half the sample frequency of the time series that is undergoing filtering. For example, a cutoff frequency for an hourly time series must be less than 12 day^{-1}. A cutoff frequency for a daily time series must be less than 2 day^{-1}.

As mentioned above, steeper frequency decreases can be achieved through using more than one filter STAGE; up to three STAGEs are allowed by TSPROC. However, if a STAGE keyword is not supplied, then a single stage is assumed. Although more stages mean greater signal rejection within the frequency stopband, the resulting propensity for "ringing," and the greater phase lag between the input and output signals, may be unwanted in many hydrologic applications.

Frequencies within the pass band of a filter are conveyed with minimal attenuation. However, as the edge of the pass band is approached, and outside the pass band, the input time series is attenuated. The decrease of output amplitude with increasing or decreasing frequency outside the pass band is referred to as "roll-off" in filtering jargon. The more stages that a filter employs, the steeper is this roll-off. Steep roll-off comes at a price—the tendency for the filter output to oscillate or "ring" in response to high amplitude events within the input time series. A phase delay between the input and output time series can also be introduced. An octave is a doubling of frequency, and a dB is a measure of signal power gain or loss. Roll-off is 6 dB/octave for a one-stage Butterworth filter, 12 dB/octave for a two-stage Butterworth filter, and 18 dB/octave for a three-stage Butterworth filter. Roll-off is 3dB/octave for one pass of the baseflow separation filter and 9 dB /octave for three passes of this filter. A roll-off rate of 6 dB/octave is equivalent to a halving of output amplitude with every factor of two changes in frequency; this is sufficient for most applications in surface-water hydrology.

Baseflow Separation Filter

Only two keywords are required to specify the characteristics of a baseflow separation filter, the ALPHA and PASSES keywords. ALPHA is the rate of decay of baseflow relative to current flow rate, and a value of 0.92– 0.98 is suitable for most applications; however, as pointed out by Nathan and McMahon (1990), a little trial and error may be required for selecting the most appropriate value for any particular application. The value of ALPHA is independent of the series sample interval. PASSES is similar to the STAGE keyword required by the Butterworth filter; yet it is also a little different in that different internal filter coefficients are not used for different passes. Furthermore, only one or three passes can be implemented, and the second pass is implemented in the reverse direction to mitigate phase shifts. If the PASSES keyword is not supplied, a value of 1 is assumed.

The output of the baseflow separation filter is a time series that represents the quickflow component of streamflow. Baseflow can then be obtained by subtracting quickflow from the original streamflow time series by use of the SERIES_EQUATION block. The occurrence of negative filtered terms, or terms that are greater than the original streamflow record, can be prevented by clipping—(see below).

Clipping (Baseflow Separation Only)

The outputs of the baseflow separation filter (but not the Butterworth filter) can be clipped to prevent negative values or values that are greater than those of the input time series. Negative values can be prevented by use of the CLIP_ZERO keyword, and values that are higher than the input time series can be prevented by use of the CLIP_INPUT keyword (fig. 11); in both cases, a "yes" or "no" specifier must be provided in the DIGITAL_FILTER block to enable clipping, and a default of "no" is assumed. The action of this filter is to provide time series that have similar characteristics to baseflow and quickflow, and this can be helpful for calibration of a model where the contribution of baseflow and quickflow to the objective function can be monitored and enhanced, if desired, through appropriate weights selection.

Settling Time

A digital filter sometimes takes a while to "settle down" when filtering operations begin on a time series; this applies more to a multistage Butterworth filter than to the baseflow separation filter implemented by TSPROC. To ensure integrity of a filtered time series, it may sometimes be necessary to remove the first part of the filtered series by use of the REDUCE_TIME_SPAN block.

Reverse Filtering

If FILTER_TYPE is set to "butterworth," FILTER_PASS is set to "low" and STAGES is set to 2, then the second stage of Butterworth filtering can be performed in the reverse direction to that of the first stage of filtering. Low pass filtering can incur a substantial phase shift, thereby delaying peaks and troughs occurring within the original time series. This phase change-induced delay can be rectified by running the digital filter from late times to early times in the second filtering stage. The user should use reverse filtering with caution, because it can sometimes actually amplify the low frequency component of filtered peaks and troughs remaining after the two-stage filtering operation has taken place.

ERASE_ENTITY

If a time series or table, especially one that contains many terms, is no longer required by TSPROC, then it can be erased from TSPROC's memory to make room for other TSPROC entities. This is achieved through use of the ERASE_ENTITY block. Keywords found in the ERASE_ENTITY block are listed in table 3. An example of an ERASE_ENTITY block is shown in figure 12. Keywords in the ERASE_ENTITY block can be supplied in any order except for the CONTEXT keyword(s), which must precede all other keywords.

Table 3.   Keywords in an ERASE_ENTITY Block.

Keyword	Role	Specifications
CONTEXT	At least one CONTEXT keyword must be supplied; up to five are permitted. If one of the CONTEXT strings matches the CONTEXT string in the SETTINGS block, or if one of the CONTEXT strings is "all," the ERASE_ENTITY block will be processed.	Any character string without internal spaces of 20 or fewer characters. The CONTEXT keyword(s) must precede all other keywords.
SERIES_NAME	Optional. The name of a time series to be erased.	A name of 18 or fewer characters referencing a time series stored within TSPROC's memory.
C_TABLE_NAME	Optional. The name of a C_TABLE to be erased.	A name of 18 characters or fewer referencing a C_TABLE stored within TSPROC's memory.
S_TABLE_NAME	Optional. The name of an S_TABLE to be erased.	A name of 18 or fewer characters referencing an S_TABLE stored within TSPROC's memory.
V_TABLE_NAME	Optional. The name of a V_TABLE to be erased.	A name of 18 or fewer characters referencing a V_TABLE stored within TSPROC's memory.
E_TABLE_NAME	Optional. The name of a E_TABLE to be erased.	A name of 18 or fewer characters referencing a E_TABLE stored within TSPROC's memory.
G_TABLE_NAME	Optional. The name of a G_TABLE to be erased.	A name of 18 or fewer characters referencing a G_TABLE stored within TSPROC's memory.

```
    START ERASE_ENTITY
     CONTEXT context_1
     C_TABLE_NAME compare
     E_TABLE_NAME ex_flow
     S_TABLE_NAME stat_flow
     V_TABLE_NAME vol_flow
     SERIES_NAME flow
    END ERASE_ENTITY
```

Figure 12.   Example of an ERASE_ENTITY block.

EXCEEDANCE_TIME

The EXCEEDANCE_TIME block instructs TSPROC to calculate the duration of time when user-supplied flows or fluxes have or have not been exceeded. The outcomes of EXCEEDANCE_TIME calculations are stored in an E_TABLE. Like every other storage entity used by TSPROC, the user must provide a name for each E_TABLE produced in this manner so that it can be referenced in later processing.

Keywords available in the EXCEEDANCE_TIME block are listed in table 4; an example of an EXCEEDANCE_TIME block is shown in figure 13. Keywords can be supplied in any order, except for the CONTEXT keyword(s),which must precede all others, and the DELAY keyword, which (if used) must directly follow a FLOW keyword.

Table 4. Keywords in an EXCEEDANCE_TIME block.

Keyword	Role	Specifications
CONTEXT	At least one CONTEXT keyword must be supplied; up to five are permitted. If one of the CONTEXT strings matches the CONTEXT string in the SET-TINGS block, or if one of the CONTEXT strings is "all," the EXCEEDANCE_TIME block will be processed.	Any character string without internal spaces of 20 or fewer characters. The CONTEXT keyword(s) must precede all other keywords.
SERIES_NAME	Mandatory. The name of the time series on which exceedance time calculations will be carried out.	A name of 18 or fewer characters referencing a time series stored within TSPROC's memory.
NEW_E_TABLE_NAME	Mandatory. The name of the new E_TABLE used to store the outcomes of exceedance time calculations carried out by TSPROC.	Any character string without spaces up to 18 characters.
EXCEEDANCE_TIME_UNITS	Mandatory. The time units to be used for storage of calculated exceedance times.	"year," "month," "day," "hour," "min," or "sec".
UNDER_OVER	Optional. Informs TSPROC whether to calculate times for which flow thresholds are exceeded ("over") or are not exceeded ("under").	"under" or "over"; default is "over".
FLOW	At least one FLOW keyword must be present. This is a threshold for which exceedance times are to be calculated.	Real number.
DELAY	Optional; but if supplied for one FLOW, must be supplied for all FLOWs. The time delay for any one event before time accumulation commences.	Real number. Time units are same as those supplied with EXCEEDANCE_TIME_UNITS keyword.

```
    START EXCEEDANCE TIME
    CONTEXT all
    SERIES_NAME outflow
    NEW E TABLE NAME et flow
    EXCEEDANCE_TIME_UNITS days
    FLOW 0.0
    FLOW 10.0
    FLOW 20.0
    FLOW 50.0
    FLOW 100.0
    FLOW 200.0
    END EXCEEDANCE_TIME
```

Figure 13. Example of an EXCEEDANCE_TIME block.

EXCEEDANCE_TIME_UNITS can be one of the following:"year" (years), "month" (months), "day" (days), "hour" (hours), "min" (minutes), "sec" (seconds). The choice of time units affects the calculated effect of the "DELAY" keyword, the calculation of the total time encompassed by the time series, and the units in which the exceedance tables are reported.

Any number of FLOW keywords can be provided in an EXCEEDANCE_TIME block. For each such FLOW, if UNDER_OVER is set to "over" (or if this keyword is omitted) and if no DELAY keywords are supplied, TSPROC calculates the time accumulated when the specified flow is exceeded. Alternatively, if UNDER_OVER is set to "under," TSPROC calculates the accumulated time when flow was less than each specified FLOW. Note that in carrying out these calculations, TSPROC does more than simply count the number of time-series terms that exceed, or are less than, the value of each FLOW, and then multiply the number of terms by the series sampling interval. This would be an incorrect procedure for two reasons. First, as mentioned above, TSPROC does not assume a uniform sampling interval for any series. Second, an exceedance-time calculation that is carried out in this way on the basis of a model-generated time series will be slightly discontinuous with respect to model parameters, which will lead to a degradation in the performance of PEST as it attempts to estimate these parameters. Instead, TSPROC carries out linear interpolation between the terms of a time series to find the "exact time" when a FLOW threshold was crossed and commences or ceases accumulating time from that point. The result is a continuous relationship between exceedance times and parameters as the latter vary during a parameter-estimation process.

Exceedance times calculated by TSPROC can be stored internally (and listed through the LIST_OUTPUT block) in time units of years, months, days, hours, minutes, or seconds. The user must set one of these options through the mandatory EXCEEDANCE_TIME_UNIT keyword.

Exceedance time calculations carried out by TSPROC need not be limited to time series that represent flow. A suitable time series could represent any environmental quantity; the numbers following the FLOW keywords in the EXCEEDANCE_TIME block would then refer to the same quantity.

Use of the DELAY keyword requires special consideration. An EXCEEDANCE_TIME block in which this keyword is featured is shown in figure 14.

```
START EXCEEDANCE_TIME
  CONTEXT all
  SERIES_NAME sim_flow
  NEW_E_TABLE_NAME sim_extime
  EXCEEDANCE_TIME_UNITS days
  UNDER_OVER under
  FLOW 20.0
  DELAY 3.0
  FLOW 50.0
  DELAY 10.0
  FLOW 100.0
  DELAY 15.0
  FLOW 200.0
  DELAY 20.0
END EXCEEDANCE_TIME
```

Figure 14.   Example of an EXCEEDANCE_TIME block showing the use of the DELAY keyword.

If a DELAY keyword is used, it must directly follow the FLOW keyword to which it pertains. Furthermore, a DELAY keyword must follow all FLOW keywords, or follow none at all.

Use of the DELAY keyword controls how exceedance time is accumulated over the period spanned by a time series. In the example shown above, UNDER_OVER is set to "under," so that time for the first FLOW entry (20 in this example) is accumulated for elements of the "sim_flow" series that have values less than 20. However, for any event where the time-series value is less than 20, time accumulation does not begin until 3 days after the beginning of the event. The time units pertaining to the DELAY keyword are assumed to be those supplied with the EXCEEDANCE_TIME_UNITS keyword.

Use of the DELAY keyword can be particularly useful when studying the effect of stream condition on biotic health. In many instances, the lethality of a particular adverse condition is a function of the magnitude and duration of the condition. The more harmful the condition, the shorter the time that elapses before the condition exerts a deleterious influence on system health. This relationship is often described by "toxicity curves" relating, for example, concentration of a constituent to the exposure time. The greater is the concentration, the less is the exposure time required to cause damage. By accumulating the time when a user-specified chemical concentration or sediment load is exceeded (or for which flow is below a user-specified threshold), and by subtracting the time required for the onset of harmful effects during each such "toxicity event," the total time when biotic health suffered can be calculated.

FLOW_DURATION

A flow-duration curve is a cumulative frequency curve that shows the percent of time values are equaled or exceeded during a given period (Searcy, 1959). The FLOW_DURATION block causes TSPROC to calculate observed or modeled values associated with a range of exceedance probabilities. FLOW_DURATION calculations are stored in a G_TABLE, and must be provided a unique name for reference later in the TSPROC instruction file.

Keywords available in the FLOW_DURATION block are listed in table 5; an example of an FLOW_DURATION block is shown in figure 15. Keywords can be supplied in any order, except for the CONTEXT keyword(s),which must precede all others. If no EXCEEDANCE_PROBABILITIES keyword is provided, values (flows) are calculated for the following exceedance probabilities: 99.5%, 99%, 98%, 95%, 90%, 75%, 50%, 25%, 10%, 5%, 2%, 1%, 0.5%. A custom probability list may be specified with the EXCEEDANCE_PROBABILITIES keyword. The list of percentages provided with this keyword may be specified in any order and are separated by spaces.

Table 5. Keywords in a FLOW_DURATION block.

[dd/mm/yyyyormm/dd/yy,where dd is the number of digits for representing the day, mm represents the month, and yyyy or yy represents the year; hh:mm:ss, where hh is the number of digits representing the hour, mm represents minutes, and ssr epresents seconds]

Keyword	Role	Specifications
CONTEXT	At least one CONTEXT keyword must be supplied; up to five are permitted. If one of the CONTEXT strings matches the CONTEXT string in the SETTINGS block, or if one of the CONTEXT strings is "all," the FLOW_DURATION block will be processed.	Any character string without internal spaces of 20 or fewer characters. The CONTEXT keyword(s) must precede all other keywords.
SERIES_NAME	Mandatory. The name of the time series on which flow duration calculations will be carried out.	A name of 18 or fewer characters referencing a time series stored within TSPROC's memory.
NEW_G_TABLE_NAME	Mandatory. The name of the new G_TABLE used to store the outcomes of flow duration calculations carried out by TSPROC.	Any character string without spaces up to 18 characters.
EXCEEDENCE_PROBABILITIES	Optional. This keyword allows for calculation of values associated with a custom list of exceedence probabilities.	Space-delimited list of integers and/or real values ranging from 0. to 100. Need not be sorted.
DATE_1	Optional. Terms of time any series before TIME_1 on this date are not imported.	Either dd/mm/yyyy or mm/dd/yy, depending on the DATE_FORMAT setting in the SETTINGS block.
TIME_1	Optional. Terms of any time series before this time on DATE_1 are not imported.	hh:mm:ss .
DATE_2	Optional. Terms of any time series after TIME_2 on this date are not imported.	Either dd/mm/yyyy or mm/dd/yy, depending on the DATE_FORMAT setting in the SETTINGS block.
TIME_2	Optional. Terms of any time series after this time on DATE_2 are not imported.	hh:mm:ss .

```
# This first block shows the usage when the default probabilities
# are accepted
START FLOW_DURATION
  CONTEXT all
  SERIES_NAME Oq6500
  NEW G TABLE NAME q dur obs
END FLOW_DURATION

# This second block shows the usage when the user provides the
# probabilities of interest
START FLOW DURATION
  CONTEXT all
  SERIES_NAME Sq6500
  NEW_G_TABLE_NAME q_dur_obs_cust
  EXCEEDANCE_PROBABILITIES 3. 7. 11. 99. 87. 66. 45. 33.
END FLOW_DURATION
```

Figure 15. Example of a FLOW_DURATION block.

If a user desires that EXCEEDANCE_TIME calculations be restricted to a certain date/time interval, a time series can be shortened prior to EXCEEDANCE_TIME calculations by use of the REDUCE_TIME_SPAN block.

GET_MUL_SERIES_GSFLOW_GAGE

The GET_MUL_SERIES_GSFLOW_GAGE block is used for extracting one or a number of time series from a "gage file" produced by the USGS GSFLOW model (Markstrom and others, 2008). Additional details about this file format are provided in Appendix 1.

Keywords available in the GET_MUL_SERIES_GSFLOW_GAGE block are listed in table 6; an example of a GET_MUL_SERIES_GSFLOW_GAGE block is shown in figure 16. Keywords may be supplied in any order, except for the CONTEXT keyword(s), which must precede all others.

Table 6.   Keywords in a GET_MUL_SERIES_GSFLOW_GAGE block.

[dd/mm/yyyy or mm/dd/yy, where dd is the number of digits for representing the day, mm represents the month, and yyyy or yy represents the year; hh:mm:ss, where hh is the number of digits representing the hour, mm represents minutes, and ss represents seconds]

Keyword	Role	Specifications
CONTEXT	At least one CONTEXT keyword must be supplied; up to five are permitted. If one of the CONTEXT strings matches the CONTEXT string in the SETTINGS block, or if one of the CONTEXT strings is "all," the GET_MUL_SERIES_GSFLOW_GAGE block will be processed.	Any character string without internal spaces of 20 or fewer characters. The CONTEXT keyword(s) must precede all other keywords.
FILE	Mandatory. The name of the GSFLOW gage file containing the time series to be imported.	Any file name up to 120 characters. Use quotes if the filename contains blank characters.
DATA_TYPE	Mandatory. The name of a data type to be imported; data types comprise column headers within a GSFLOW gage file. The name of a new series must immediately follow the DATA_TYPE keyword in the GET_MUL_SERIES_GSFLOW_GAGE block.	Any character string without internal spaces up to 30 characters.
NEW_SERIES_NAME	Mandatory. The name of a new series as stored by TSPROC. This keyword must immediately follow a DATA_TYPE keyword, which specifies the dataset imported from the site sample file.	Any character string without internal spaces up to 18 characters.
MODEL_REFERENCE_DATE	Mandatory. The date corresponding to zero model simulation time. Simulation times are recorded in the GSFLOW gage file.	Either dd/mm/yyyy or mm/dd/yy, depending on the DATE_FORMAT setting in the SETTINGS block.
MODEL_REFERENCE_TIME	Mandatory. The time corresponding to zero model simulation time. Simulation times are recorded on the GSFLOW gage file.	Either dd/mm/yyyy or mm/dd/yy, depending on the DATE_FORMAT setting in the SETTINGS block.
TIME_UNITS_PER_DAY	Optional. The number of time units as employed by the GSFLOW gage file comprising one day. If omitted, this is assumed to be 1.	A real number.
DATE_1	Optional. Terms of time any series before TIME_1 on this date are not imported.	Either dd/mm/yyyy or mm/dd/yy, depending on the DATE_FORMAT setting in the SETTINGS block.
TIME_1	Optional. Terms of any time series before this time on DATE_1 are not imported.	hh:mm:ss .
DATE_2	Optional. Terms of any time series after TIME_2 on this date are not imported.	Either dd/mm/yyyy or mm/dd/yy, depending on the DATE_FORMAT setting in the SETTINGS block.
TIME_2	Optional. Terms of any time series after this time on DATE_2 are not imported.	hh:mm:ss.

```
      START GET_MUL_SERIES_GSFLOW_GAGE
       CONTEXT all
       FILE file1a.ggo
       DATA_TYPE flow
       NEW_SERIES_NAME flow_s
       DATA_TYPE uzf-runoff
       NEW_SERIES_NAME uzfr_s
       TIME_UNITS_PER_DAY 1
       MODEL_REFERENCE_DATE 1/1/2000
       MODEL_REFERENCE_TIME 12:00:00
       DATE_1 3/4/2005
       TIME_1 12:00:00
       DATE_2 6/9/2010
       TIME_2 00:00:00
      END GET_MUL_SERIES_GSFLOW_GAGE
```

Figure 16. Example of a GET_MUL_SERIES_GSFLOW_GAGE block.

Entries pertaining to each imported series must be grouped; thus, one or a number of pairs of DATA_TYPE and NEW_SERIES_NAME keywords (in that order) must be provided in the GET_MUL_SERIES_GSFLOW_GAGE block. DATA_TYPE refers to a column header in the GSFLOW gage file, while NEW_SERIES_NAME provides the name of the imported series stored by TSPROC. The number of these keyword pairs that can appear in a GET_MUL_SERIES_GSFLOW_GAGE block is not limited; a new series is imported for each such pair.

Each GSFLOW gage file contains a "time" column designated by an appropriate header. Entries in this column are normally in days; TSPROC uses days as the default unless a TIME_UNITS_PER_DAY keyword is present within the GET_MUL_SERIES_GFLOW_GAGE block. However, where the time units are different from days, TSPROC must employ a time conversion factor as it imports the time series, and this factor is supplied as the entry following the TIME_UNITS_PER_DAY keyword. Suppose that time units are actually hours; then TIME_UNITS_PER_DAY should be supplied as 24.0. Thus, as the name suggests, the number of time units employed by the model collectively comprise 1 day.

To convert model simulation time to days and times, a reference date and time are needed, and they are the date and time corresponding to a simulation time of zero. These items must be supplied following the MODEL_REFERENCE_DATE and MODEL_REFERENCE_TIME keywords, which are mandatory in the GET_MUL_SERIES_GSFLOW_GAGE block.

The DATE_1, TIME_1 and DATE_2, TIME_2 keywords can be employed to restrict the length of the time series that is imported into TSPROC. No entries before DATE_1, TIME_1 or after DATE_2, TIME_2 are imported. Missing TIME_1 and TIME_2 entries denote a time of 00:00:00 in each case. Either or both of the DATE_1 and DATE_2 keywords can be omitted from the GET_MUL_SERIES_GSFLOW_GAGE block. If both of them are missing, the entire time series is imported.

GET_MUL_SERIES_SSF / GET_SERIES_SSF

The GET_MUL_SERIES_SSF block allows one or more series to be imported from a single site sample file in one operation. The format of the site sample file is generic and simple (Appendix 1); it can be used, for example, to store processed observation data for later use in a lengthy parameter-estimation run. Note that for compatibility with previous versions of TSPROC, a "GET_SERIES_SSF" block is processed as described in this section as well.

Table 7 shows the keywords appearing in a GET_MUL_SERIES_SSF block. An example of a GET_MUL_SERIES_SSF block is shown in figure 17.

Table 7. Keywords in a GET_MUL_SERIES_SSF block.

[dd/mm/yyyy or mm/dd/yy, where dd is the number of digits for representing the day, mm represents the month, and yyyy or yy represents the year; hh:mm:ss, where hh is the number of digits representing the hour, mm represents minutes, and ss represents seconds]

Keyword	Role	Specifications
CONTEXT	At least one CONTEXT keyword must be supplied; up to five are permitted. If one of the CONTEXT strings matches the CONTEXT string in the SETTINGS block, or if one of the CONTEXT strings is "all," the GET_MUL_SERIES_SSF block will be processed.	Any character string without internal spaces of 20 or fewer characters. The CONTEXT keyword(s) must precede all other keywords.
FILE	Mandatory. The name of the site sample file containing the time series to be imported.	Any file name up to 120 characters. Use quotes if the filename contains blank characters.
SITE	Mandatory. The name of a site within the site sample file for which a time series is to be imported. The name of a new series must immediately follow the SITE keyword in the GET_MUL_SERIES_SSF block.	Any character string without internal spaces up to 18 characters.
NEW_SERIES_NAME	Mandatory. The name of a new series as stored by TSPROC. This keyword must immediately follow the SITE keyword which specifies the dataset imported from the site sample file.	Any character string without internal spaces up to 18 characters.
DATE_1	Optional. Terms of time series before TIME_1 on this date are not imported.	Either dd/mm/yyyy or mm/dd/yy, depending on the DATE_FORMAT setting in the SETTINGS block.
TIME_1	Optional. Terms of time series before this time on DATE_1 are not imported.	hh:mm:ss .
DATE_2	Optional. Terms of time series after TIME_2 on this date are not imported.	Either dd/mm/yyyy or mm/dd/yy, depending on the DATE_FORMAT setting in the SETTINGS block.
TIME_2	Optional. Terms of time series after this time on DATE_2 are not imported.	hh:mm:ss .

```
     START GET_MUL_SERIES_SSF
      CONTEXT all
      FILE flows.ssf
      SITE rebec_ck
      NEW_SERIES_NAME rebecca
      SITE horton_ck
      NEW_SERIES_NAME horton
      SITE sandy_ck
      NEW_SERIES_NAME sandy
      DATE_1 06/03/1970
      TIME_1 12:00:00
      DATE_2 09/01/1980
      TIME_2 00:00:00
     END GET_MUL_SERIES_SSF
```

Figure 17. Example of a GET_MUL_SERIES_SSF block.

The DATE_ and TIME_ specifiers are optional. If they are omitted, then the entire time series pertaining to each of the nominated sites is imported. If a DATE_1 keyword is present but a TIME_1 keyword is absent, then TIME_1 is assumed to be 00:00:00; this also applies for DATE_2. If TIME_1 is present, then DATE_1 must be present; the same holds for TIME_2.

A GET_MUL_SERIES_SSF block can contain multiple instances of the SITE and NEW_SERIES_NAME keywords. These keywords must be supplied in pairs, with the SITE keyword immediately preceding the associated NEW_SERIES_NAME keyword. The same site cannot be supplied twice.

Correct operation of the instructions contained within the GET_MUL_SERIES_SSF block assumes that the site sample file read by use of this block is correct and consistent. This includes the date format of the site sample file, which must be consistent with date format specified in the SETTING block of the TSPROC input file.

GET_MUL_SERIES_STATVAR

STATVAR files are written by the USGS Precipitation-Runoff Modeling System (PRMS) model (Leavesley and others, 1983), as well as by the USGS GSFLOW model (Markstrom and others, 2008). An example of such a file is given in figure 18.

```
4
node_cfs.musroute 132
runoff.obs 16
node_cfs.musroute 87
node_cfs.musroute 14
1 1975 6 1 0 0 0 2.490942 6.434562 3.300000 0.000000
2 1975 6 2 0 0 0 2.501948 7.389743 2.800000 0.000000
3 1975 6 3 0 0 0 2.476184 9.652343 2.900000 0.000000
{…continues…}
```

Figure 18. Example of the header and first three lines of a STATVAR file.

Keywords that can be employed in a GET_MUL_SERIES_STATVAR block are provided in table 8. An example of a GET_MUL_SERIES_STATVAR block is given in figure 19.

Table 8. Keywords in a GET_MUL_SERIES_STATVAR block.

[dd/mm/yyyy or mm/dd/yy, where dd is the number of digits for representing the day, mm represents the month, and yyyy or yy represents the year; hh:mm:ss, where hh is the number of digits representing the hour, mm represents minutes, and ss represents seconds]

Keyword	Role	Specifications
CONTEXT	At least one CONTEXT keyword must be supplied; up to five are permitted. If one of the CONTEXT strings matches the CONTEXT string in the SETTINGS block, or if one of the CONTEXT strings is "all," the GET_MUL_SERIES_STATVAR block will be processed.	Any character string without internal spaces of 20 or fewer characters. The CONTEXT keyword(s) must precede all other keywords.
FILE	Mandatory. The name of the STATVAR file containing the time series to be imported.	Any file name up to 120 characters. Use quotes if the filename contains blank characters.
VARIABLE_NAME	Mandatory. The name of a variable within the STATVAR file for which a time series is to be imported. Collectively a variable name and location id denote a unique time series.	Any character string without internal spaces up to 50 characters.
LOCATION_ID	Mandatory. The location identifier for an imported time series. The location id together with a variable name collectively denote a unique time series. This keyword must immediately follow a VARIABLE_NAME keyword.	An integer.
NEW_SERIES_NAME	Mandatory. The name of a new series as stored by TSPROC. This keyword must immediately follow a LOCATION_ID keyword.	Any character string without internal spaces up to 18 characters.
DATE_1	Optional. Terms of any time series before TIME_1 on this date are not imported.	Either dd/mm/yyyy or mm/dd/yy, depending on the DATE_FORMAT setting in the SETTINGS block.
TIME_1	Optional. Terms of any time series before this time on DATE_1 are not imported.	hh:mm:ss .
DATE_2	Optional. Terms of any time series after TIME_2 on this date are not imported.	Either dd/mm/yyyy or mm/dd/yy, depending on the DATE_FORMAT setting in the SETTINGS block.
TIME_2	Optional. Terms of any time series after this time on DATE_2 are not imported.	hh:mm:ss .

```
START GET_MUL_SERIES_STATVAR
 CONTEXT all
 FILE statvar.dat
 VARIABLE_NAME node_cfs.musroute
 LOCATION_ID 132
 NEW_SERIES_NAME runoff16
 VARIABLE_NAME node_cfs.musroute
 LOCATION_ID 2
 NEW_SERIES_NAME cfs2
 DATE_1 6/4/1975
 TIME_1 12:00:00
 DATE_2 10/29/2001
 TIME_2 00:00:00
END GET_ MUL_SERIES_STATVAR
```

Figure 19.   Example of a GET_MUL_SERIES_STATVAR block.

A GET_MUL_SERIES_STATVAR block can be used to import one or a number of series from a STATVAR file. Each imported series is identified by a VARIABLE_NAME and LOCATION_ID. These two keywords must be supplied in that order, immediately followed by a NEW_SERIES_NAME keyword for each series to be imported. As many such triplets must be featured in this block as there are series to import.

The DATE_1, TIME_1, DATE_2 and TIME_2 keywords are optional. If none of these keywords is supplied, then the entirety of each time series is imported. If any of these keywords are present, then no series terms that predate DATE_1, TIME_1 or postdate DATE_2, TIME_2 are imported. If TIME_1 or TIME_2 is omitted, a time of 00:00:00 is assumed in either case.

GET_MUL_SERIES_PLOTGEN / GET_SERIES_PLOTGEN

The GET_MUL_SERIES_PLOTGEN block governs importation of time-series data from a HSPF PLOTGEN file into TSPROC. Each NEW_SERIES_NAME keyword in this block must directly follow a LABEL keyword to clarify the association between the time-series label in the HSPF PLOTGEN file and the name of the new series stored within TSPROC. Table 9 lists the allowable keywords in a GET_MUL_SERIES_PLOTGEN block. For compatibility with previous versions of TSPROC, a control file that contains a "GET_SERIES_PLOTGEN" block is recognized and treated as a GET_MUL_SERIES_PLOTGEN block. Figure 20 shows an example of a GET_MUL_SERIES_PLOTGEN block.

Table 9. Keywords in a GET_MUL_SERIES_PLOTGEN block.

[dd/mm/yyyy or mm/dd/yy, where dd is the number of digits for representing the day, mm represents the month, and yyyy or yy represents the year; hh:mm:ss, where hh is the number of digits representing the hour, mm represents minutes, and ss represents seconds]

Keyword	Role	Specifications
CONTEXT	At least one CONTEXT keyword must be supplied; up to five are permitted. If one of the CONTEXT strings matches the CONTEXT string in the SETTINGS block, or if one of the CONTEXT strings is "all," the GET_MUL_SERIES_PLOTGEN block will be processed.	Any character string without internal spaces of 20 or fewer characters. The CONTEXT keyword(s) must precede all other keywords.
FILE	Mandatory. The name of the PLOTGEN file containing the time series to be imported.	Any file name up to 120 characters. Use quotes if the filename contains spaces.
LABEL	At least one LABEL keyword must be present. This is the PLOTGEN label pertaining to a time series that is to be imported.	Any character string up to 20 characters. If the string contains blank characters, enclose it in quotes.
NEW_SERIES_NAME	Mandatory. The name of the new series as stored by TSPROC. This must immediately follow the LABEL keyword pertaining to the imported time series.	Any character string without spaces up to 18 characters.
DATE_1	Optional. Terms of the time series before TIME_1 on this date are not imported.	Either dd/mm/yyyy or mm/dd/yyyy, depending on the DATE_FORMAT setting in the SETTINGS block.
TIME_1	Optional. Terms of the time series before this time on DATE_1 are not imported.	hh:mm:ss .
DATE_2	Optional. Terms of the time series after TIME_2 on this date are not imported.	Either dd/mm/yyyy or mm/dd/yy, depending on the DATE_FORMAT setting in the SETTINGS block.
TIME_2	Optional. Terms of the time series after this time on DATE_2 are not imported.	hh:mm:ss .

```
      START GET_MUL_SERIES_PLOTGEN
      CONTEXT all
      FILE hspfout.plt
      LABEL "total outflow"
      NEW_SERIES_NAME t_outflow
      LABEL interflow
      NEW_SERIES_NAME interflow
      DATE_1 6/1/1976
      TIME_1 00:12:00
      DATE_2 7/1/1976
      TIME_2 00:12:00
      END GET_MUL_SERIES_PLOTGEN
```

Figure 20. Example of a GET_MUL_SERIES_PLOTGEN block.

DATE_ and TIME_ specifiers are optional in a GET_MUL_SERIES_PLOTGEN block. If they are absent from the block, then the entire time series pertaining to each nominated label is imported. If a DATE_1 keyword is present but a TIME_1 keyword is absent, then TIME_1 is assumed to be 00:00:00; the same applies for DATE_2. However, if TIME_1 is present, then DATE_1 must also be present; the same holds for TIME_2.

GET_SERIES_WDM

Instructions provided in this block allow TSPROC to import a time series from a Watershed Data Management (WDM) file. Many hydrologic and water-quality models and analyses developed by the U.S. Geological Survey and the U.S. Environmental Protection Agency currently use a WDM file. The WDM file is a binary file that provides the user with a common data base for many applications, thus eliminating the need to reformat data from one application to another.

Table 10 shows the keywords permissible in a GET_SERIES_WDM block. An example of a GET_SERIES_WDM block is given in figure 21.

Table 10. Keywords within a GET_SERIES_WDM block.

[dd/mm/yyyy or mm/dd/yy, where dd is the number of digits for representing the day, mm represents the month, and yyyy or yy represents the year; hh:mm:ss, where hh is the number of digits representing the hour, mm represents minutes, and ss represents seconds]

Keyword	Role	Specifications
CONTEXT	At least one CONTEXT keyword must be supplied; up to five are permitted. If one of the CONTEXT strings matches the CONTEXT string in the SETTINGS block, or if one of the context strings is "all," the GET_SERIES_WDM block will be processed.	Any string without internal spaces of 20 or fewer characters. The CONTEXT keyword(s) must precede all other keywords.
FILE	Mandatory. The name of the WDM file containing the time series to be imported.	Any file name up to 120 characters. Use quotes if the filename contains blank characters.
NEW_SERIES_NAME	Mandatory. The name of the new time series as stored by TSPROC.	Any character string without internal spaces up to 18 characters.
DSN	Mandatory. The data set number of the time series to be imported.	Any integer for which a time series dataset is available within the nominated WDM file.
DATE_1	Optional. Terms of the time series before TIME_1 on this date are not imported.	Either dd/mm/yyyy or mm/dd/yy, depending on the DATE_FORMAT setting in the SETTINGS block.
TIME_1	Optional. Terms of the time series before this time on DATE_1 are not imported.	hh:mm:ss .
DATE_2	Optional. Terms of the time series after TIME_2 on this date are not imported.	Either dd/mm/yyyy or mm/dd/yy, depending on the DATE_FORMAT setting in the SETTINGS block.
TIME_2	Optional. Terms of the time series after this time on DATE_2 are not imported.	hh:mm:ss .
DEF_TIME	Optional. If the time series imported from a WDM file has a time step of 1 day or greater, each term pertains to a date, but not to a time. Upon importation into TSPROC, each term of such a series is referenced to REF_TIME. Default is 00:00:00.	hh:mm:ss. If DEF_TIME is supplied as 24:00:00, a time of 00:00:00 on the following day will be assigned to all samples.
FILTER	Optional. Terms of the time series that have this value are ignored upon importation into TSPROC.	Real number.

```
    START GET_SERIES_WDM
     CONTEXT all
     FILE catchment.wdm
     DSN 1013
     NEW SERIES NAME coal ck
     DATE_1 06/03/1970
     TIME_1 12:00:00
     DATE_2 09/01/1980
     TIME_22 00:00:00
     DEF_TIME 12:00:00
     FILTER -999.99
    END GET_SERIES_WDM
```

Figure 21.   Example usage of a GET_SERIES_WDM block.

The DATE_ and TIME_ specifiers are optional in a GET_SERIES_WDM block. If they are omitted, then the entire time series pertaining to the nominated dataset number is imported. If a DATE_1 keyword is present but a TIME_1 keyword is absent, then TIME_1 is assumed to be 00:00:00,and a similar assumption is made for DATE_2. If TIME_1 is present, then DATE_1 must be present; the same holds for TIME_2.

If the sample interval for a time series stored in a WDM file is 1 day or greater, then each term of the series will have no time reference; however, each time series term within TSPROC is associated with both a date and a time. When such a time series is imported from a WDM file into TSPROC, TSPROC's default behavior is to assign each term a time of 00:00:00 on the day with which it is associated. This time can be altered to the user's choice by use of the optional DEF_TIME keyword. Note that if DEF_TIME is supplied as "24:00:00," then each sample is assigned a time of 00:00:00 on the following day.

HYDRO_EVENTS

The HYDRO_EVENTS block allows storm hydrographs to be extracted for the days preceding and following peak flow events; events may defined by specifying a minimum number of days between peaks and by assigning a minimum peak value. The terms of the series upon which events are extracted may be limited to those events within a specified date/time interval.

TSPROC stores the outcomes of storm events extracted by the HYDRO_EVENTS block in a new time series. Like other TSPROC entities, the new time series must be provided with a name so that it can be referenced by other TSPROC processing blocks. This name must be 18 or fewer characters and must not include a space character.

Keywords featured in the HYDRO_EVENTS block are listed in table 11. An example of a HYDRO_EVENTS block is given in figure 22.

Table 11.   Keywords in a HYDRO_EVENTS block.

[dd/mm/yyyy or mm/dd/yy, where dd is the number of digits for representing the day, mm represents the month, and yyyy or yy represents the year; hh:mm:ss, where hh is the number of digits representing the hour, mm represents minutes, and ss represents seconds]

Keyword	Role	Specifications
CONTEXT	At least one CONTEXT keyword must be supplied; up to five are permitted. If one of the CONTEXT strings matches the CONTEXT string in the SETTINGS block, or if one of the CONTEXT strings is "all," the PERIOD_STATISTICS block will be processed.	Any string without internal spaces of 20 or fewer characters. The CONTEXT keyword(s) must precede all other keywords.
SERIES_NAME	Mandatory. The name of the time series on which statistical calculations will be carried out.	A name of 18 or fewer characters referencing a time series stored within TSPROC's memory.
NEW_SERIES_NAME	Mandatory. The name of the new time series used to store the outcomes of peak extractions.	Any character string without internal spaces up to 18 characters.
WINDOW	Optional. Minimum time between successive peaks, in days.	A real number other than zero. Default is 1.
MIN_PEAK	Optional. Minimum value for a peak.	A nonnegative real number. Default is 0.
RISE_LAG	Mandatory. The number of days prior to the peak to include in the event hydrograph.	A nonzero real number.
FALL_LAG	Mandatory. The number of days following the peak to include in the event hydrograph.	A nonzero real number.
DATE_1	Optional. Terms of the time series before TIME_1 on this date are not used in peak extraction.	Either dd/mm/yyyy or mm/dd/yy, depending on the DATE_FORMAT setting in the SETTINGS block.
TIME_1	Optional. Terms of the time series before this time on DATE_1 are not used in peak extraction.	hh:mm:ss .
DATE_2	Optional. Terms of the time series after TIME_2 on this date are not used in peak extraction.	Either dd/mm/yyyy or mm/dd/yy, depending on the DATE_FORMAT setting in the SETTINGS block.
TIME_2	Optional. Terms of the time series after this time on DATE_2 are not used in peak extraction.	hh:mm:ss .

```
    START HYDRO_EVENTS
     CONTEXT all
     SERIES_NAME outflow
     NEW_SERIES_NAME outflow_pk
     WINDOW 7
     MIN_PEAK 10.5
     RISE_LAG 2
     FALL_LAG 6
     DATE_1 10/1/1976
     TIME_1 00:00:00
     DATE_2 9/30/1985
     TIME_2 00:00:00
    END HYDRO_EVENTS
```

Figure 22.   Example of a HYDRO_EVENTS block.

HYDROLOGIC_INDICES

Hydrologic indices are statistical measures applied to streamflow records to quantify various ecologically important aspects of a flow regime. Classes of hydrologic indices include measures that quantify such things as the timing and seasonal pattern of extreme flows; daily, seasonal, and annual flow variability; and the rates of flow increases and decreases. Over 160 different hydrologic indices may be calculated by means of a HYDROLOGIC_INDICES block. The indices are computed by means of code adapted from the USGS Hydrologic Index Tool (Henricksen and others, 2006). A complete list of all indices that may be calculated in a HYDROLOGIC_INDICES block is included in Appendix 3.

Many of the hydrologic indices are, in fact, the mean of other summary statistics. For example, index value MH20 is the specific *mean* annual maximum flow. TSPROC can also calculate these indices as median values of the summary statistics. The keyword USE_MEDIAN may be inserted into the HYDROLOGIC_INDICES block to cause all such indices to be calculated as median , rather than mean, values. If USE_MEDIAN is present in the HYDROLOGIC_INDICES block, the index MH20 then would assume the definition "specific *median* annual maximum flow." Appendix 3 notes all hydrologic indices that are affected by the USE_MEDIAN keyword.

The list of hydrologic indices to be calculated may be specified in one of two ways: 1) by means of the STREAM_CLASSIFICATION and FLOW_COMPONENT keywords, or 2) through use of the two-letter keywords (for example, "MA"), followed by the index number (as listed in Appendix 3).

Table 12 lists the keywords available for use in a HYDROLOGIC_INDICES block. Note that all indices are calculated and reported if the user does not supply STREAM_CLASSIFICATION or FLOW_COMPONENT keywords and arguments and do not specify individual hydrologic indices.

The two-letter keywords accept any number of space-delimited values as arguments. For example, the following keyword and arguments could be used to specify that the indices associated with magnitude of average flows, numbers 12, 22, and 25 are calculated: "MA 12 22 25" (see Appendix 3 for definitions of these indices). The possible argument values that may be specified for the STREAM_CLASSIFICATION and FLOW_COMPONENT keywords are shown in table 13.

Because many of the hydrologic indices that can be calculated are somewhat redundant, the user may elect to calculate a subset of the available indices on the basis of the stream's hydrologic regime type (for example, snowmelt dominated, groundwater dominated, surface runoff dominated). Olden and Poff (2003) identified a list of the high information, nonredundant indices for six stream classification types. Table 14 summarizes the arguments that may be supplied to the STREAM_CLASSIFICATION and FLOW_COMPONENT keywords to generate a subset of hydrologic indices appropriate to the system under study. The hydrologic index definitions referenced in table 14 are from Olden and Poff (2003).

For example, specifying "STREAM_CLASSIFICATION snowmelt_perennial" and "FLOW_COMPONENT high_flow_frequency" would result in the calculation and reporting of just two indices: FH_8 and FH_{11}. Subsequent FLOW_COMPONENT entries append the appropriate indices to the list to be calculated. Multiple instances of both the STREAM_CLASSIFICATION and the FLOW_COMPONENT keywords may be present in the control file. For example, as shown in figure 23 and table 14, a STREAM_CLASSIFICATION value of "snow_rain_perennial" and a second STREAM_CLASSIFICATION value of "snowmelt_perennial" is used with the FLOW_COMPONENT values "timing," "average_magnitude," and "low_flow_magnitude" to select indices MA3, MA29, MA40, MA44, ML13, ML14, ML22, TA1, TA3, and TL1 for calculation. If only the STREAM_CLASSIFICATION or FLOW_COMPONENT keyword is given (not both), this has the effect of selecting all indices identified by Olden and Poff (2003) associated with the given stream classification or flow component. In other words, specifying "STREAM_CLASSIFICATION flashy_intermittent" results in calculation of the indices associated with a flashy intermittent stream type for all flow components.

The direct specification of indices ("FL 1 2 3") in the block shown in figure 23 would cause TSPROC to add the indices "FL1," "FL2," and "FL3" to the list of indices in the preceding paragraph. The results are added to a new G_TABLE (general table), which may be manipulated or output just like any other TSPROC table.

Note that not all of the indices TSPROC can calculate are suited to optimization runs involving PEST; all derivatives in PEST must be smooth and continuous in order for optimization to succeed. Recall that PEST evaluates the model response to small perturbations in parameter values. Indices involving counts will not produce smooth derivatives. The indices expected to result in poor derivatives and thus poor PEST optimization results include the frequency indices (FL, FH), some of the duration indices (DL18–number of zero-flow days; DL19–variability in number of low-flow days; DL20–percentage of all months with zero flow) and some of the timing indices (TL1–Julian date of annual minimum; TH1–Julian data of annual maximum).

Table 12. Keywords in a HYDROLOGIC_INDICES block.

[dd/mm/yyyy or mm/dd/yy, where dd is the number of digits for representing the day, mm represents the month, and yyyy or yy represents the year; hh:mm:ss, where hh is the number of digits representing the hour, mm represents minutes, and ss represents seconds]

Keyword	Role	Specifications
CONTEXT	At least one CONTEXT keyword must be supplied; up to five are permitted. If one of the CONTEXT strings matches the CONTEXT string in the SETTINGS block, or if one of the CONTEXT strings is "all," the HYDROLOGIC_INDICES block will be processed.	Any string without internal spaces of 20 or fewer characters. The CONTEXT keyword(s) must precede all other keywords.
SERIES_NAME	Mandatory. The name of the time series on which statistical calculations will be carried out.	Any file name up to 120 characters. Use quotes if the filename contains blank characters.
NEW_G_TABLE_NAME -or- NEW_TABLE_NAME	Mandatory. The name of the new time series used to store the outcomes of hydrologic index calculations.	Any character string without internal spaces up to 18 characters.
USE_MEDIAN	Optional. Requests that indices that normally report the mean of some other sumamry statistic to instead report the median value.	
STREAM_CLASSIFICATION	Optional. Specify the hydrologic regime as defined in Olden and Poff (2003).	See table 13 for list of legal values.
FLOW_COMPONENT	Optional. Specify the hydrologic regime as defined in Olden and Poff (2003).	See table 13 for list of legal values.
MA	Optional. Magnitude—average-flow conditions.	Valid range: 1–45
ML	Optional. Magnitude—low-flow conditions.	Valid range: 1–22
MH	Optional. Magnitude—high-flow conditions.	Valid range: 1–27
FL	Optional. Frequency—low-flow conditions.	Valid range: 1–3
FH	Optional. Frequency—high-flow conditions.	Valid range: 1–10
DL	Optional. Duration—low-flow conditions.	Valid range: 1–20
DH	Optional. Duration—high-flow conditions.	Valid range: 1–22
TA	Optional. Timing—average-flow conditions.	Valid range: 1–3
TL	Optional. Timing—low-flow conditions.	Valid range: 1–4
TH	Optional. Timing—high-low conditions.	Valid range: 1–3
RA	Optional. Rate of change—average-flow conditions.	Valid range: 1–9
DATE_1	Optional. Terms of the time series before TIME_1 on this date are not used in calculation of indices.	Either dd/mm/yyyy or mm/dd/yy depending on the DATE_FORMAT setting in the SETTINGS block.
TIME_1	Optional. Terms of the time series before this time on DATE_1 are not used in calculation of indices.	hh:mm:ss .
DATE_2	Optional. Terms of the time series after TIME_2 on this date are not used in calculation of indices.	Either dd/mm/yyyy or mm/dd/yy, depending on the DATE_FORMAT setting in the SETTINGS block.
TIME_2	Optional. Terms of the time series after this time on DATE_2 are not used in calculation of indices.	hh:mm:ss .

Table 13. Argument values for the STREAM_CLASSIFICATION and FLOW_COMPONENT keywords.

Keyword	Legal argument values
STREAM_CLASSIFICATION	HARSH_INTERMITTENT
	FLASHY_INTERMITTENT
	SNOWMELT_PERENNIAL
	SNOW_RAIN_PERENNIAL
	GROUNDWATER_PERENNIAL
	FLASHY_PERENNIAL
	ALL_STREAMS
FLOW_COMPONENT	AVERAGE_MAGNITUDE
	LOW_FLOW_MAGNITUDE
	HIGH_FLOW_MAGNITUDE
	LOW_FLOW_FREQUENCY
	HIGH_FLOW_FREQUENCY
	LOW_FLOW_DURATION
	HIGH_FLOW_DURATION
	TIMING
	RATE_OF_CHANGE

```
    START HYDROLOGIC_INDICES
     CONTEXT all
     SERIES_NAME outflow
     NEW_TABLE_NAME outflow_hi
     USE MEDIAN
     STREAM_CLASSIFICATION snow_rain_perennial
     FLOW_COMPONENT timing
     FLOW_COMPONENT low_flow_magnitude
     STREAM_CLASSIFICATION snowmelt_perennial
     FLOW_COMPONENT average_magnitude
     FL 1 2 3
     DATE_1 01/01/1980
     TIME_1 00:00:00
     DATE_2 12/31/1989
     TIME_2 23:59:59
    END HYDROLOGIC_INDICES
```

Figure 23. Example of a HYDROLOGIC_INDICES block.

Table 14. Matrix of possible hydrologic indices selected by combining FLOW_COMPONENT and STREAM_CLASSIFICATION arguments.

FLOW_COMPONENT argument	Stream classification and flow component from Olden and Poff (2003)	STREAM_CLASSIFICATION argument						
		HARSH_INTERMITTENT	FLASHY_INTERMITTENT	SNOWMELT_PERENNIAL	SNOW_RAIN_PERENNIAL	GROUNDWATER_PERENNIAL	FLASHY_PERENNIAL	ALL_STREAMS
		Harsh intermittent	Intermittent flashy or runoff	Perenial snowmelt	Perennial snow or rain	Perennial superstable or stable groundwater	Perennial flashy or runoff	All streams
AVERAGE_MAGNITUDE	Magnitude of flow events (average-flow conditions)	MA34, MA22, MA16	MA37, MA18, MA21, MA9	MA29, MA40	MA3, MA44	MA3, MA41, MA8	MA26, MA41, MA10	MA5, MA41, MA3, MA11
LOW_FLOW_MAGNITUDE	Magnitude of flow events (low-flow conditions)	ML13, ML15, ML1	ML16, ML6, ML22, ML15	ML13, ML22	ML13, ML14	ML18, ML14, ML16	ML17, ML14, ML16	ML17, ML4, ML21, ML18
HIGH_FLOW_MAGNITUDE	Magnitude of flow events (high-flow conditions)	MH23, MH14, MH9	MH23, MH4, MH14, MH7	MH1, MH20	MH17, MH20	MH17, MH19, MH10	MH23, MH8, MH14	MH16, MH8, MH10, MH14
LOW_FLOW_FREQUENCY	Frequency of low-flow events	FL2, FL3, FL1	FL2, FL3, FL1	FL2, FL3	FL2, FL3	FL2, FL3, FL1	FL2, FL3	FL2, FL3, FL1
HIGH_FLOW_FREQUENCY	Frequency of high-flow events	FH2, FH5, FH7	FH2, FH3, FH7, FH10	FH8, FH11	FH3, FH5	FH3, FH6, FH11	FH4, FH6, FH7	FH2, FH3, FH6, FH7
LOW_FLOW_DURATION	Duration of low-flow events	DL1, DL2, DL13	DL1, DL13, DL16, DL18	DL5, DL16	DL6, DL13	DL9, DL11, DL16	DL6, DL10, DL17	DL13, DL16, DL17, DL18
HIGH_FLOW_DURATION	Duration of high-flow events	DH5, DH10, DH22	DH12, DH13, DH15, DH23	DH16, DH19	DH12, DH24	DH11, DH15, DH20	DH13, DH16, DH24	DH13, DH15, DH16, DH20
TIMING	Timing of flow events	TH1, TL2, TH2	TA1, TA2, TL1, TH3	TA1, TA3	TA1, TL1	TA1, TH1, TL1	TA1, TA3, TH3	TA1, TH3, TL2
RATE_OF_CHANGE	Rate of change in flow events	RA4, RA1, RA5	RA9, RA6, RA5, RA7	RA1, RA8	RA9, RA8	RA9, RA8, RA5	RA9, RA7, RA6	RA9, RA8, RA6, RA5

LIST_OUTPUT

The LIST_OUTPUT block causes the series and tables generated by TSPROC to be written to an ASCII text list output file or site sample file. An instruction file by which PEST can read the contents of a list output file can be generated automatically by use of the WRITE_PEST_FILES block.

Keywords associated with a LIST_OUTPUT block are given in table 15. An example of a LIST_OUTPUT block is shown in figure 24. Keywords can be supplied in any order, except for the CONTEXT keyword(s), which must precede all others.

Table 15.   Keywords in a LIST_OUTPUT block.

Keyword	Role	Specifications
CONTEXT	At least one CONTEXT keyword must be supplied; up to five are permitted. If one of the CONTEXT strings matches the CONTEXT string in the SETTINGS block, or if one of the CONTEXT strings is "all," the LIST_ OUTPUT block will be processed.	Any string without internal spaces of 20 or fewer characters. The CONTEXT keyword(s) must precede all other keywords.
FILE	Mandatory. The name of the file to be written by the LIST_OUTPUT block.	Any filename up to 120 characters. Use quotes if the filename contains spaces.
SERIES_NAME	Optional. The name of a time series to be written by the LIST_OUTPUT block to its output file.	A name of 18 or fewer characters referencing a time series stored within TSPROC's memory.
SERIES_FORMAT	Mandatory if a SERIES_NAME keyword is present. Determines whether dates, times and the series name should accompany series terms in the file generated by the LIST_OUTPUT block.	"short," "long," or "ssf".
C_TABLE_NAME	Optional. The name of a c_TABLE to be written by the LIST_OUTPUT block to its output file.	A name of 18 or fewer characters referencing a C_TABLE stored within TSPROC's memory.
S_TABLE_NAME	Optional. The name of an S_TABLE to be written by the LIST_OUTPUT block to its output file.	A name of 18 or fewer characters referencing an S_TABLE stored within TSPROC's memory.
V_TABLE_NAME	Optional. The name of a V_TABLE to be written by the LIST_OUTPUT block to its output file.	A name of 18 or fewer characters referencing a V_TABLE stored within TSPROC's memory.
E_TABLE_NAME	Optional. The name of an E_TABLE to be written by the LIST_OUTPUT block to its output file.	A name of 18 or fewer characters referencing an E_TABLE stored within TSPROC's memory.
G_TABLE_NAME	Optional. The name of a G_TABLE to be written by the LIST_OUTPUT block to its output file.	A name of 18 or fewer characters referencing a G_TABLE stored within TSPROC's memory.

```
    START LIST_OUTPUT
    CONTEXT all
    FILE output.txt
    SERIES_NAME flow_216
    SERIES_NAME flow_342
    S_TABLE_NAME st_216
    S_TABLE_NAME st_342
    C_TABLE_NAME comp_ser
    V_TABLE_NAME vol_216
    V_TABLE_NAME vol_342
    E_TABLE_NAME dur_216
    E_TABLE_NAME dur_342
    SERIES_FORMAT short
    END LIST_OUTPUT
```

Figure 24.   Example of a LIST_OUTPUT block.

Any number of time series or tables can be written to a file generated by the LIST_OUTPUT block; as many of the keywords regarding these entities as desired can be supplied in this block. When the block generates its output files, time series are written first, followed (in order) by S_TABLEs, C_TABLEs, V_TABLEs, E_TABLEs, and finally G_TABLEs. The order of the individual entities of each type within the different segments of the TSPROC output file is the same as the order in which respective keywords referencing those entities is supplied in the LIST_OUTPUT block.

If a SERIES_NAME keyword is provided in a LIST_OUTPUT block, then a SERIES_FORMAT keyword must also be provided; options are "short," "long," and "ssf." If "short" is specified, the LIST_OUTPUT block lists the terms of the time series as a single column in its output file. If the "long" option is specified, the terms of the time series are accompanied by the date and time corresponding to the term, as well as the name of the time series. This "long" format corresponds to that of a site sample file (see Appendix 1), and the output file can thus be used by other members of the PEST Surface Water Utilities after the header to each time series is removed. If SERIES_FORMAT is specified as "ssf," TSPROC eliminates the header for each time series, and the user is spared having to manually remove the header of a list output file.

If TSPROC is run as part of a composite model under the control of PEST, it is best to use the "short" option for time-series formatting. This is because, where a time series is large, a considerable amount of computation time may be spent converting TSPROC's internal representation of sample dates and times to the dd/mm/yyyy (or mm/dd/yyyy) and hh:mm:ss formats required for output listing, and this can add considerably to overall composite model execution time. Note also that if the "long" protocol is employed, then in accordance with site sample file protocol, TSPROC does not represent midnight as "24:00:00"; instead midnight is represented as 00:00:00 on the following day.

Output formatting for other TSPROC entities is such that they are clearly labeled and easily understood by the user. In the case of S_TABLEs, G_TABLEs and C_TABLEs, it is important to note that only the statistics that have been calculated in prior blocks can be written in the LIST_OUTPUT block.

Exceedance times stored in an E_TABLE are recorded by the LIST_OUTPUT block as accumulated times and as proportions of the total time spanned by the parent time series. If this file is used by PEST, then only the exceedance proportions are actually read by PEST on the basis of the instruction file created through a WRITE_PEST_FILES block.

NEW_SERIES_UNIFORM

The NEW_SERIES_UNIFORM block creates a uniform-valued time series with series entries placed at uniform (or almost uniform) time intervals. Keywords belonging to the NEW_SERIES_UNIFORM block are listed in table 16. An example NEW_SERIES_UNIFORM block is given in figure 25

Table 16. Keywords in a NEW_SERIES_UNIFORM block.

Keyword	Role	Specifications
CONTEXT	At least one CONTEXT keyword must be supplied; up to five are permitted. If one of the CONTEXT strings matches the CONTEXT string in the SETTINGS block, or if one of the CONTEXT strings is "all," the NEW_SERIES_UNIFORM block will be processed.	Any string without internal spaces of 20 or fewer characters. The CONTEXT keyword(s) must precede all other keywords.
NEW_SERIES_NAME	Mandatory. The name of the new, uniform-valued time series.	Any character string without internal spaces up to 18 characters.
DATE_1	Mandatory. The starting date of the new time series..	Either dd/mm/yyyy or mm/dd/yy, depending on the DATE_FORMAT in the SETTINGS block.
TIME_1	Mandatory. The starting time of the new time series..	hh:mm:ss .
DATE_2	Mandatory. Terms of the new time series are not created after this date.	Either dd/mm/yyyy or mm/dd/yy, depending on the DATE_FORMAT in the SETTINGS block.
TIME_2	Mandatory. Terms of the new time series are not created after this time on DATE_2.	hh:mm:ss .
TIME_INTERVAL	Mandatory. The number of TIME_UNITS (see below) between terms of the new series.	An integer greater than zero.
TIME UNIT	Mandatory. The time units in which the TIME_INTER-VAL is expressed.	"seconds," "minutes," "hours," "days," "months" or "years" .
NEW_SERIES_VALUE	Mandatory. The value supplied to all new terms in the new time series.	A real number.

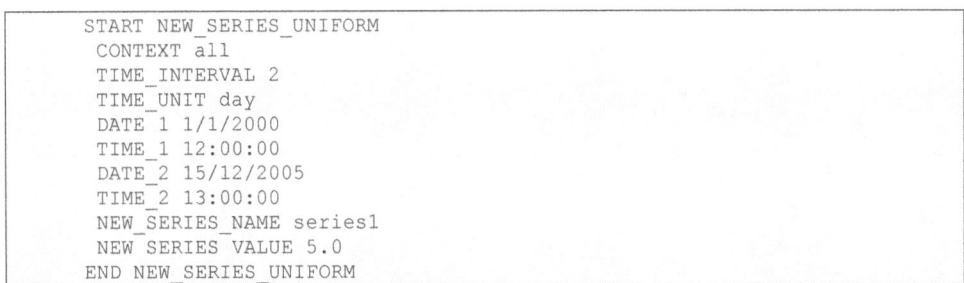

```
     START NEW_SERIES_UNIFORM
      CONTEXT all
      TIME_INTERVAL 2
      TIME_UNIT day
      DATE_1 1/1/2000
      TIME_1 12:00:00
      DATE_2 15/12/2005
      TIME_2 13:00:00
      NEW_SERIES_NAME series1
      NEW_SERIES_VALUE 5.0
     END NEW_SERIES_UNIFORM
```

Figure 25. Example of a NEW_SERIES_UNIFORM block.

The behavior of the NEW_SERIES_UNIFORM block is slightly different depending on whether TIME_UNIT is supplied as "second," "minute," "hour," or "day" on the one hand, or "month" or "year" on the other hand. In the former case, the series time interval is strictly TIME_INTERVAL times the TIME_UNIT; thus, the time increment between successive terms of the series is strictly uniform. For example, if the TIME_INTERVAL is supplied as 2 and the TIME_UNIT is supplied as "hours," then all samples are 2 hours apart. The first sample occurs on DATE_1, TIME_1; the last sample is no later than DATE_2, TIME_2. (Note that, in contrast to most other blocks, all of the DATE_1, TIME_1, DATE_2, and TIME_2 keywords must be supplied.)

If TIME_UNIT is set to "month," then all terms of the new series occur on the same day of the month and at the same time as the initial time TIME_1. Thus, terms are variously 28, 29, 30, or 31 days apart. (In this case, TSPROC rejects a DATE_1 in which the date is the 29th, 30th, or 31st of the month.) If TIME_UNIT is set to "year," then all terms of the new time series occur on the same date, but they are separated by TIME_INTERVAL years. In this case, TSPROC does not allow DATE_1 to be the 29th of February. All terms of the new series are assigned the same user-supplied NEW_SERIES_VALUE.

The NEW_TIME_SERIES block can be useful as a precursor to digital filtering; digital filtering can take place only on a time series whose terms are separated by a constant time increment. If filtering must be performed on an observed time series that has values present at an irregular time increment, the series can be interpolated to a constant time base series by:

4. creating a series with uniform time increments by means of a NEW_SERIES_UNIFORM block; and

5. creating a new series (with regular time intervals and interpolated values) by means of the NEW_TIME_BASE block, specifying the series created in step 1 above as the new time base.

The interpolated time series can then be filtered.

NEW_TIME_BASE

The NEW_TIME_BASE block is used to interpolate values from the sample times pertaining to one time series to the sample times pertaining to another. Keywords belonging to the NEW_TIME_BASE block are listed in table 17. An example NEW_TIME_BASE block is shown in figure 26.

Table 17.   Keywords in a NEW_TIME_BASE block.

Keyword	Role	Specifications
CONTEXT	At least one CONTEXT keyword must be supplied; up to five are permitted. If one of the CONTEXT strings matches the CONTEXT string in the SETTINGS block, or if one of the CONTEXT strings is "all," the NEW_TIME_BASE block will be processed.	Any string without internal spaces of 20 or fewer characters. The CONTEXT keyword(s) must precede all other keywords.
SERIES_NAME	Mandatory. The name of the time series whose terms are to be time interpolated to a new time base.	A name of 18 or fewer characters referencing a time series stored within TSPROC's memory.
NEW_SERIES_NAME	Mandatory. The name of the new time series produced as an outcome of time interpolation.	Any character string without internal spaces up to 18 characters.
TB_SERIES_NAME	Mandatory. The name of the time series to whose dates and times time interpolation will take place.	A name of 18 or fewer characters referencing a time series stored within TSPROC's memory.

```
      START NEW_TIME_BASE
       CONTEXT all
       SERIES_NAME mod_flow
       TB_SERIES_NAME obs_flow
       NEW_SERIES_NAME int_flow
      END NEW_TIME_BASE
```

Figure 26.   Example of a NEW_TIME_BASE block.

One time series, specified by SERIES_NAME, will be time interpolated to the time base of another, specified by TB_SERIES_NAME, only if the time span of the series TB_SERIES_NAME is equal to or smaller than that of the series SERIES_NAME. The result of the interpolation is a new time series specified by NEW_SERIES_NAME, with terms that have exactly the same dates and times as those of the time-base time series. If the original time series and the time-base time series pertain to the same data type, time interpolation allows the two series to be directly compared with each other. This "apples with apples" comparison is crucial when calibrating a model against field data. Hence, one of the principal roles of TSPROC, when used as a model post-processor in a "composite model" run by PEST, is to carry out this important time interpolation of model-generated time series to the dates and times of their measured counterparts. This interpolated time series can then be written to a TSPROC output file (using the LIST_OUTPUT block), where it can be read by PEST and compared with measured values recorded in a PEST control file. Both the PEST control file and the instruction file by which the time-interpolated time series can be read from the LIST_OUTPUT file can be written by use of the WRITE_PEST_FILES block.

PERIOD_STATISTICS

The PERIOD_STATISTICS block is similar to the SERIES_STATISTICS block except for one key difference: treatment of dates. In a SERIES_STATISTICS block, all date ranges must be specified explicitly and included in an external dates file. In a PERIOD_STATISTICS block, the date ranges are calculated internally for the more common date ranges, obviating the need for an external dates file. By use of the PERIOD_STATISTICS block, a number of simple statistics can be calculated for specific periods within a time series. Optionally, the terms of the series upon which statistical calculations are based can be limited to those within a specified date/time interval. Another option provided by the PERIOD_STATISTICS block is for statistics to be calculated on the basis of the log (to base 10) of the terms of the time series or on the terms of the series raised to an arbitrary power. If it is desired that statistics be calculated on the basis of more complex functions of the terms of a time series, this can be easily achieved by first calculating a new time series by use of the SERIES_EQUATION block, and then undertaking statistical calculations on the basis of this new time series.

At present, seven statistical measures can be calculated using the PERIOD_STATISTICS block: these include the mean, standard deviation, median, sum, maximum, minimum, and range. The statistics may be calculated for a monthly series (for example, "12/1997," "01/1998," "02/1998") or monthly mean (for example, the mean for all values in the month of January for the included time period).

TSPROC stores the outcomes of statistical calculations carried out by the PERIOD_STATISTICS block in a new time series. As for other TSPROC entities, the new time series must be provided with a name so that it can be referenced by other TSPROC processing blocks. This name must be 18 or fewer characters and must not include a space character.

Keywords featured in the PERIOD_STATISTICS block are listed in table 18. An example of a PERIOD_STATISTICS block is shown in figure 27.

Table 18.   Keywords in a PERIOD_STATISTICS block.

[dd/mm/yyyy or mm/dd/yy, where dd is the number of digits for representing the day, mm represents the month, and yyyy or yy represents the year; hh:mm:ss, where hh is the number of digits representing the hour, mm represents minutes, and ss represents seconds]

Keyword	Role	Specifications
CONTEXT	At least one CONTEXT keyword must be supplied; up to five are permitted. If one of the CONTEXT strings matches the CONTEXT string in the SETTINGS block, or if one of the CONTEXT strings is "all," the PERIOD_STATISTICS block will be processed.	Any string without internal spaces of 20 or fewer characters. The CONTEXT keyword(s) must precede all other keywords.
SERIES_NAME	Mandatory. The name of the time series on which statistical calculations will be carried out.	A name of 18 or fewer characters referencing a time series stored within TSPROC's memory.
NEW_SERIES_NAME	Mandatory. The name of the new time series used to store the outcomes of statistical calculations.	Any character string without internal spaces up to 18 characters.
STATISTIC	Mandatory. Key word describing the statistic to be computed.	"mean," "std_dev," "median," "sum," "maximum," "minimum," or "range"
PERIOD	Mandatory. Key word describing the period for which statistics are to be computed.	"month_many," "month_one," or "year".
YEAR_TYPE	Optional when PERIOD is "YEAR." Describes the type of year to be used for each period.	"water_high," "water_low," or "calendar" Default is "water_high."
TIME_ABSCISSA	Mandatory. Informs TSPROC whether the date and time corresponding to each new time series entry pertains to the beginning, middle, or end of the corresponding period.	"start," "centre," "center," or "end".
LOG	Optional. Requests that statistics be calculated based on the logs (to base 10) of the terms of the time series.	"yes" or "no." Default is "no." The LOG keyword cannot be used if the POWER keyword is used.
POWER	Optional. Requests that statistics be calculated based on the terms of the time series raised to the nominated power.	A real number other than zero. Default is 1. The POWER keyword cannot be used if the LOG keyword is used.
DATE_1	Optional. Terms of the time series before TIME_1 on this date are not used in statistics calculations.	Either dd/mm/yyyy or mm/dd/yy, depending on the DATE_FORMAT setting in the SETTINGS block.
TIME_1	Optional. Terms of the time series before this time on DATE_1 are not used in statistical calculations.	hh:mm:ss .
DATE_2	Optional. Terms of the time series after TIME_2 on this date are not used in statistics calculations.	Either dd/mm/yyyy or mm/dd/yy, depending on the DATE_FORMAT setting in the SETTINGS block.
TIME_2	Optional. Terms of the time series after this time on DATE_2 are not used in statistical calculations.	hh:mm:ss .

```
   START PERIOD STATISTICS
    CONTEXT all
    SERIES_NAME outflow
    NEW_SERIES_NAME outflow_m
    STATISTIC mean
    PERIOD month_many
    TIME ABSCISSA start
    LOG yes
    DATE_1 10/1/1976
    TIME_1 00:00:00
    DATE_2 9/30/1985
    TIME_2 00:00:00
   END PERIOD_STATISTICS
```

Figure 27. Example of a PERIOD_STATISTICS block.

The use of the POWER and LOG keywords may have unintended consequences. It is illegal for both of these keywords to be present within the same PERIOD_STATISTICS block. Furthermore, using these keywords may cause numerical errors. In particular, if LOG is set to "yes" and if any of the terms of the time series are zero or negative, TSPROC ceases execution and displays an error message. In addition, if a POWER with an absolute value of less than 1 is supplied and if any of the terms of the time series are negative, or if the POWER is negative and any of the terms of the time series are zero, TSPROC likewise ceases execution and displays an error message instead of attempting this impossible calculation.

The traditional or high-flow water year begins October 1 of the previous year and ends September 30 of an individual year. The low-flow water year begins April 1 of the previous year and ends March 31 of an individual year (Riggs, 1972). The calendar year begins January 1 and ends December 31 of an individual year.

For monthly statistics, selecting "period_many" as the PERIOD keyword computes a monthly statistic for each year in the input time series. Alternatively, selecting "period_one" as the PERIOD keyword computes a single statistic for each month, aggregating across all years in the input time series. When computing a mean, the "period_many" mean is commonly called the monthly mean value, whereas the "period_one" mean is commonly called the mean monthly value.

REDUCE_TIME_SPAN

The REDUCE_TIME_SPAN block reduces the time spanned by a time series, and this may be a useful precursor to other steps in TSPROC processing. For example, by use of the REDUCE_TIME_SPAN block, the time spanned by an observed time series can be reduced to that spanned by a model-generated time series. This allows time interpolation of the model's output times to the observation times to be carried out using the NEW_TIME_BASE block.

Keywords found in a REDUCE_TIME_SPAN block are listed in table 19. An example of a REDUCE_TIME_SPAN block is shown in figure 28.

Table 19.   Keywords in a REDUCE_TIME_SPAN block.

[dd/mm/yyyy or mm/dd/yy, where dd is the number of digits for representing the day, mm represents the month, and yyyy or yy represents the year; hh:mm:ss, where hh is the number of digits representing the hour, mm represents minutes, and ss represents seconds]

Keyword	Role	Specifications
CONTEXT	At least one CONTEXT keyword must be supplied; up to five are permitted. If one of the CONTEXT strings matches the CONTEXT string in the SETTINGS block, or if one of the CONTEXT strings is "all," the REDUCE_TIME_SPAN block will be processed.	Any string without internal spaces of 20 or fewer characters. The CONTEXT keyword(s) must precede all other keywords.
SERIES_NAME	Mandatory. The name of the time series whose time span is to be reduced.	A name of 18 or fewer characters referencing a time series stored within TSPROC's memory.
NEW_SERIES_NAME	Mandatory. The name of the new time series produced as an outcome of time span reduction.	Any character string without internal spaces up to 18 characters.
DATE_1	Optional. Terms of the time series before TIME_1 on this date are not copied to the new time series.	Either dd/mm/yyyy or mm/dd/yy, depending on the DATE_FORMAT in the SETTINGS block.
TIME_1	Optional. Terms of the time series before this time on DATE_1 are not copied to the new time series.	hh:mm:ss .
DATE_2	Optional. Terms of the time series after TIME_2 on this date are not copied to the new time series.	Either dd/mm/yyyy or mm/dd/yy, depending on the DATE_FORMAT in the SETTINGS block.
TIME_2	Optional. Terms of the time series after this time on DATE_2 are not copied to the new time series.	hh:mm:ss .

```
    START REDUCE_TIME_SPAN
    CONTEXT all
    SERIES_NAME intflow
    NEW_SERIES_NAME intflow_1
    DATE_1 02/01/1976
    TIME_1 13:13:00
    DATE_2 06/01/1976
    TIME_2 00:00:00
    END REDUCE_TIME_SPAN
```

Figure 28.   Example of a REDUCE_TIME_SPAN block.

When a new time series is created by reducing the time span of an existing time series, the original time series still remains within TSPROC's memory. If desired, it can be removed by use of the ERASE_ENTITY block.

At least one DATE_1 or DATE_2 keyword must be supplied. If the corresponding TIME_ keyword is not supplied, a default time of 00:00:00 is used. If the DATE_1 keyword is omitted, DATE_1 and TIME_1 are assumed to be the first date and time cited in the original time series; in other words, no time-span reduction from the front of the time series takes place. Similarly, if the DATE_2 keyword is omitted, no time-span reduction takes place from the end of the existing time series. Note that a TIME_ keyword cannot be supplied without the corresponding DATE_ keyword.

SERIES_BASE_LEVEL

The SERIES_BASE_LEVEL block allows a user to subtract a constant amount from all terms of a time series. This constant amount is the value of one term of an existing time series, either the time series from which subtraction is taking place or another time series stored within the memory of TSPROC.

A common use of the SERIES_BASE_LEVEL block is for generating a new time series comprised of the deviations of an existing time series relative to some base level. In this case, the first term of the time series may be taken as the base level, because this term is subtracted from all other elements of the time series to create the new series with altered base level. SERIES_BASE_LEVEL functionality allows this new series to either replace the original time series or to exist as its own separate entity.

Keywords found in a SERIES_BASE_LEVEL block are listed in table 20. An example of a SERIES_BASE_LEVEL block is given in figure 29.

Table 20.   Keywords in a SERIES_BASE_LEVEL block.

[dd/mm/yyyy or mm/dd/yy, where dd is the number of digits for representing the day, mm represents the month, and yyyy or yy represents the year; hh:mm:ss, where hh is the number of digits representing the hour, mm represents minutes, and ss represents seconds]

Keyword	Role	Specifications
CONTEXT	At least one CONTEXT keyword must be supplied; up to five are permitted. If one of the CONTEXT strings matches the CONTEXT string in the SETTINGS block, or if one of the CONTEXT strings is "all," the SERIES_BASE_LEVEL block will be processed.	Any string without internal spaces of 20 or fewer characters. The CONTEXT keyword(s) must precede all other keywords.
SERIES_NAME	Mandatory. The name of the time series whose base level is to be altered.	A name of 18 or fewer characters referencing a time series stored within TSPROC's memory.
SUBSTITUTE	Mandatory. If this is supplied as "yes," the new time series replaces the old one in TSPROC's memory and retains the same name. If it is supplied as "no," a new series is created.	"yes" or "no" .
NEGATE	Optional. If this is supplied as "yes," all terms of the new base series are multiplied by −1 after subtraction of the constant.	"yes" or "no" .
NEW_SERIES_NAME	Mandatory if SUBSTITUTE is supplied as "yes." The name of the new time series produced as an outcome of base level alteration.	Any character string without internal spaces up to 18 characters.
BASE_LEVEL_SERIES_NAME	Mandatory. The name of the time series of which one element will be subtracted from all elements of the original time series to effect the base level change.	A name of 18 or fewer characters referencing a time series stored within TSPROC's memory.
BASE_LEVEL_DATE	Mandatory. This is used with BASE_LEVEL_TIME to identify the term of series BASE_LEVEL_SERIES_NAME that is subtracted from all elements of SERIES to effect the base level change.	Either dd/mm/yyyy or mm/dd/yy, depending on the DATE_FORMAT in the SETTINGS block.
BASE_LEVEL_TIME	Mandatory. This is used with BASE_LEVEL_DATE to identify the term of series BASE_LEVEL_SERIES_NAME that is subtracted from all elements of SERIES to effect the base level change.	hh:mm:ss .

```
   START SERIES BASE LEVEL
    CONTEXT all
    SERIES_NAME head
    BASE_LEVEL_SERIES_NAME head
    BASE_LEVEL_DATE 01/04/1996
    BASE_LEVEL_TIME 12:00:00
    SUBSTITUTE no
    NEGATE yes
    NEW_SERIES_NAME drawdown
   END SERIES_BASE_LEVEL
```

Figure 29. Example of a SERIES_BASE_LEVEL block.

The SERIES_EQUATION block can also be used to subtract a constant from the terms of a series. In that case, the constant is supplied as a number in an equation. In the case of the SERIES_BASE_LEVEL block, the subtractor is a term in a series, identified through the name of the series and the date and time to which the term pertains. If there is no term corresponding to the supplied date and time, TSPROC displays an appropriate error message and ceases execution.

The NEGATE keyword reverses the sign of the output time series; this can be useful, for example, when calculating drawdown from a time series of head observations. Drawdown is calculated as the negative of the change in head from its initial value. Thus, after base level alteration by subtraction of the initial series term, all terms of the new time series are multiplied by -1.

SERIES_CLEAN

The SERIES_CLEAN block is used to eliminate unwanted terms from a time series or replace them with a preferred value. This is sometimes required for correcting the deleterious effects of model misbehavior whereby model-generated time-series are "polluted" with intermittent spurious values. It can also be used for eliminating outliers in an observation time series.

Keywords pertaining to the SERIES_CLEAN block are listed in table 21. An example SERIES_CLEAN block is shown in figure 30.

Table 21. Keywords within a SERIES_CLEAN block.

Keyword	Role	Specifications
CONTEXT	At least one CONTEXT keyword must be supplied; up to five are permitted. If one of the CONTEXT strings matches the CONTEXT string in the SETTINGS block, or if one of the CONTEXT strings is "all," the SERIES_CLEAN block will be processed.	Any string without internal spaces of 20 or fewer characters. The CONTEXT keyword(s) must precede all other keywords.
SERIES_NAME	Mandatory. The name of the time series whose terms are to be "cleaned."	A name of 18 or fewer characters referencing a time series stored within TSPROC's memory.
NEW_SERIES_NAME	Mandatory if SUBSTITUTE_VALUE is "delete"; optional otherwise. The name of a new time series formed by removal or replacement of terms in the original time series.	Any character string without internal spaces up to 18 characters.
LOWER_ERASE_BOUNDARY	Optional. Terms equal to and above this threshold and equal to and below the UPPER_ERASE_BOUNDARY threshold are removed or replaced.	A real number.
UPPER_ERASE_BOUNDARY	Optional. Terms equal to or below this threshold and equal to or above the LOWER_ERASE_BOUNDARY threshold are removed or replaced.	A real number.
SUBSTITUTE_VALUE	Mandatory. If supplied as a real number, this is the value substituted for terms between the upper and lower erase thresholds. If supplied as "delete," instructs TSPROC to delete terms between these thresholds.	A real number or "delete."

```
START SERIES_CLEAN
  CONTEXT all
  SERIES_NAME series1
  LOWER_ERASE_BOUNDARY 0.0
  UPPER_ERASE_BOUNDARY 0.03
  SUBSTITUTE_VALUE 0.0
  NEW_SERIES_NAME series2
END SERIES_CLEAN
```

Figure 30. Example of a SERIES_CLEAN block.

The SERIES_CLEAN block presents the user with different options for handling unwanted terms. In the simplest case, these terms are replaced by the number supplied through the SUBSTITUTE_VALUE keyword. If this is done, terms can be replaced "in situ" (that is, in the existing time series without creating a new one), or a new time series can be created to hold the altered time series while the original time series remains intact. If a NEW_SERIES_NAME keyword is supplied, the latter option is taken; if not, the former option is taken.

Unwanted terms can be completely removed by supplying the string "delete," instead of a real number, with the SUBSTITUTE_VALUE keyword.

Terms of a series are identified for deletion or replacement by use of the LOWER_ERASE_BOUNDARY and UPPER_ERASE_BOUNDARY keywords. Either one or both of these keywords can be supplied. If both of them are supplied, all terms of the time series between and including the specified boundary values are replaced or deleted. If only the LOWER_ERASE_BOUNDARY keyword is supplied, all terms equal to and above this threshold are removed or replaced. If only the UPPER_ERASE_BOUNDARY keyword is supplied, all terms equal to or below this boundary are removed or replaced. If you are in any doubt of the action of the SERIES_CLEAN block when only one of these keywords is supplied, then supply both of them, with one of them either very high or very low. Note that if you do this, then TSPROC does not accept numbers whose absolute value is greater than about 1.0E+37.

SERIES_COMPARE

The SERIES_COMPARE block calculates statistics that quantify the similarity of one time series with another. The outcomes of these calculations are placed in a C_TABLE, which can be written to a file using the LIST_OUTPUT block.

Keywords pertaining to the SERIES_COMPARE block are listed in table 22. An example SERIES_COMPARE block is given in figure 31.

Table 22. Keywords within a SERIES_COMPARE block.—Continued

[dd/mm/yyyy or mm/dd/yy, where dd is the number of digits for representing the day, mm represents the month, and yyyy or yy represents the year; hh:mm:ss, where hh is the number of digits representing the hour, mm represents minutes, and ss represents seconds]

Keyword	Role	Specifications
CONTEXT	At least one CONTEXT keyword must be supplied; up to five are permitted. If one of the CONTEXT strings matches the CONTEXT string in the SETTINGS block, or if one of the CONTEXT strings is "all," the SERIES_COMPARE block will be processed.	Any string without internal spaces of 20 or fewer characters. The CONTEXT keyword(s) must precede all other keywords.
SERIES_NAME_SIM	Mandatory. The name of the "simulated" time series whose terms are to be compared with the "observed" time series.	A name of 18 or fewer characters referencing a time series stored within TSPROC's memory.
SERIES_NAME_OBS	Mandatory. The name of the "observed" time series whose terms are to be compared with the "simulated" time series.	A name of 18 or fewer characters referencing a time series stored within TSPROC's memory.
SERIES_NAME_BASE	Optional if either the COEFFICIENT_OF_EFFICIENCY or INDEX_OF_AGREEMENT keyword is present. The name of a baseline time series that can be used in the calculation of these two quantities.	A name of 18 or fewer characters referencing a time series stored within TSPROC's memory.
NEW_C_TABLE_NAME	Mandatory. The name of the new C_TABLE used to store the outcomes of comparison statistics calculations.	Any character string without internal spaces up to 18 characters.
BIAS	Optional. Requests calculation of bias between the observed and simulated time series (B in the equations shown in the text).	"yes" or "no" Default is "no."
STANDARD_ERROR	Optional. Requests calculation of the standard error between the observed and simulated time series (S in the equations shown in the text).	"yes" or "no" Default is "no."
RELATIVE_BIAS	Optional. Requests calculation of the relative bias between the observed and simulated time series (Br in the equations shown in the text).	"yes" or "no" Default is "no."
RELATIVE_STANDARD_ERROR	Optional. Requests calculation of the relative standard error between the observed and simulated time series (Sr in the equations shown in the text).	"yes" or "no" Default is "no."
NASH_SUTCLIFFE	Optional. Requests calculation of the Nash-Sutcliffe (1970) coefficient (R2 in the equations shown in the text).	"yes" or "no" Default is "no."
COEFFICIENT_OF_EFFICIENCY	Optional. Requests calculation of the coefficient of efficiency (E in the equations shown in the text); see Legates and McCabe (1999).	"yes" or "no" Default is "no."
INDEX_OF_AGREEMENT	Optional. Requests calculation of the index of agreement (d in the equations shown in the text); see Legates and McCabe (1999).	"yes" or "no" Default is "no."

Table 22. Keywords within a SERIES_COMPARE block.—Continued

[dd/mm/yyyy or mm/dd/yy, where dd is the number of digits for representing the day, mm represents the month, and yyyy or yy represents the year; hh:mm:ss, where hh is the number of digits representing the hour, mm represents minutes, and ss represents seconds]

Keyword	Role	Specifications
VOLUMETRIC_EFFICIENCY	Optional. Requests calculation of a 'volumetric efficiency' metric as proposed by Criss and Winston (2008).	"yes" or "no" Default is "no."
EXPONENT	Mandatory if either the COEFFICIENT_OF_EFFICIENCY or INDEX_OF_AGREEMENT keyword is present. The exponent used in the calculation of these quantities (k in the equations shown in the text).	An integer—must be 1 or 2.
DATE_1	Optional. Terms of the simulated and observed time series before TIME_1 on this date are not used in series comparison.	Either dd/mm/yyyy or mm/dd/yy, depending on the DATE_FORMAT setting in the SETTINGS block.
TIME_1	Optional. Terms of the simulated and observed time series before this time on DATE_1 are not used in series comparison.	hh:mm:ss .
DATE_2	Optional. Terms of the simulated and observed time series after TIME_2 on this date are not used in series comparison.	Either dd/mm/yyyy or mm/dd/yy, depending on the DATE_FORMAT setting in the SETTINGS block.
TIME_2	Optional. Terms of the simulated and observed time series after this time on DATE_2 are not used in series comparison.	hh:mm:ss .

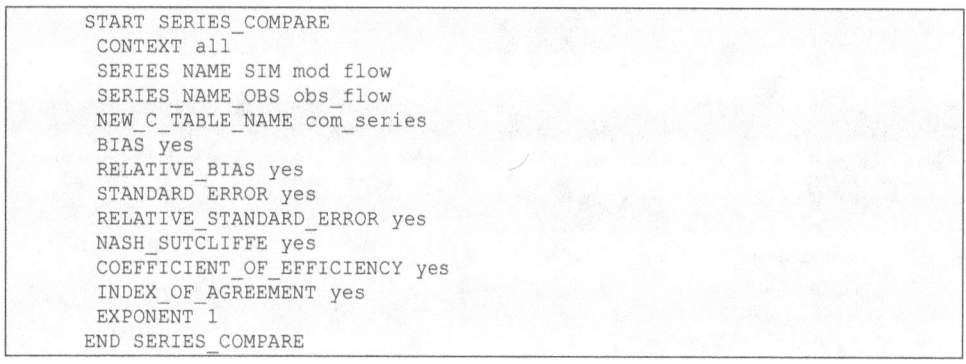

```
START SERIES_COMPARE
  CONTEXT all
  SERIES_NAME_SIM mod_flow
  SERIES_NAME_OBS obs_flow
  NEW_C_TABLE_NAME com_series
  BIAS yes
  RELATIVE_BIAS yes
  STANDARD_ERROR yes
  RELATIVE_STANDARD_ERROR yes
  NASH_SUTCLIFFE yes
  COEFFICIENT_OF_EFFICIENCY yes
  INDEX_OF_AGREEMENT yes
  EXPONENT 1
END SERIES_COMPARE
```

Figure 31. Example of a SERIES_COMPARE block.

The names of two time series must be provided in a SERIES_COMPARE block. One of the names is denoted as the "observed" time series and the other is the "simulated" time series; the name difference is important in calculating relative bias, relative standard error, the Nash-Sutcliffe coefficient (Nash and Sutcliffe, 1970), the index of agreement, and the coefficient of efficiency, because standardization of these quantities is undertaken with respect to the observed time series. The simulated and observed time series must contain samples taken at identical dates and times within the time interval spanned by the DATE_1, TIME_1 and DATE_2, TIME_2 entries. If these keywords are not provided, the sample dates and times of the observed and simulated time series must be identical over the entire length of these series.

If either of the COEFFICIENT_OF_EFFICIENCY or INDEX_OF_AGREEMENT keywords are present, then an EXPONENT keyword must be present and optionally, a SERIES_NAME_BASE keyword may be supplied to provide the name of a "baseline time series" that can be used in place of the mean observation value over the comparison time window (see Legates and McCabe (1999) for details). The theory underpinning use of the coefficient of efficiency and index of agreement as bases for series comparison is discussed in Legates and McCabe (1999). The value supplied with the keyword EXPONENT must be either 1 or 2. The baseline time series must have terms at identical dates and times to those of the simulated and observed time series over the comparison time window. If the SERIES_NAME_BASE keyword is omitted, then the mean observation is employed in the formulae presented below instead of the terms of the baseline time series.

Equations for the quantities calculated in the SERIES_COMPARE block are given below. The symbols S_i and O_i refer to individual values in a simulated and observed series, respectively. Note that the Nash-Sutcliffe coefficient is equal to the coefficient of efficiency when the exponent in the latter equation is two and when a baseline time series is not provided.

Bias:

$$B = \frac{1}{N} \Sigma \left(S_i - O_i \right) \tag{1}$$

Standard error:

$$S = \sqrt{\frac{1}{N-1} \Sigma \left(S_i - O_i \right)^2} \tag{2}$$

Relative bias:

$$B_r = \frac{B}{\overline{O}} \tag{3}$$

Relative standard error:

$$S_r = \frac{S}{S_o} \tag{4}$$

Volumetric efficiency

$$VE = 1 - \frac{\Sigma \left| S_i - O_i \right|}{\Sigma O_i} \tag{5}$$

Nash-Sutcliffe coefficient:

$$R^2 = 1 - \frac{\Sigma \left(S_i - O_i \right)^2}{\Sigma \left(O_i - \overline{O} \right)^2} \tag{6}$$

Coefficient of efficiency:

$$E_k = 1 - \frac{\Sigma \left| S_i - O_i \right|^k}{\Sigma \left| O_i - \overline{O} \right|^k} \tag{7}$$

Index of agreement:

$$d_k = 1 - \frac{\sum |S_i - O_i|^k}{\sum \left(|S_i - \bar{O}| + |O_i - \bar{O}| \right)^k}$$
(8)

where:

$$\bar{O} = \frac{1}{N} \sum O_i$$
(9)

$$S_0 = \sqrt{\frac{1}{N-1} \sum \left(O_i - \bar{O} \right)^2}$$
(10)

and N is the number of terms in the series (or subseries) between which comparison takes place; summation in the above equations takes place over all of these terms. Where a SERIES_NAME_BASE keyword is supplied, $\bar{O}$ in the equations for coefficient of efficiency and index of agreement is replaced by B_i, the respective term of the baseline time series.

If the user desires that weights be applied to terms of the series before comparison (as is often the case), weighted observation and simulated time series can easily be generated by use of the SERIES_EQUATION block.

SERIES_DIFFERENCE

The SERIES_DIFFERENCE block is used to calculate a new time series, whose terms are differences between successive terms of an existing time series. This establishes a rate-of-change time series. Keywords pertaining to the SERIES_DIFFERENCE block are listed in table 23. An example SERIES_DIFFERENCE block is given in figure 32.

Table 23. Keywords within a SERIES_DIFFERENCE block.

Keyword	Role	Specifications
CONTEXT	At least one CONTEXT keyword must be supplied; up to five are permitted. If one of the CONTEXT strings matches the CONTEXT string in the SETTINGS block, or if one of the CONTEXT strings is "all," the SERIES_DIFFERENCE block will be processed.	Any string without internal spaces of 20 or fewer characters. The CONTEXT keyword(s) must precede all other keywords.
SERIES_NAME	Mandatory. The name of the time series whose terms are to be differenced.	A name of 18 or fewer characters referencing a time series stored within TSPROC's memory.
NEW_SERIES_NAME	Mandatory. The name of the new time series formed by subtracting subsequent terms of an existing series.	Any character string without internal spaces up to 18 characters.

```
    START SERIES_DIFFERENCE
     CONTEXT all
     SERIES_NAME 322
     NEW_SERIES_NAME 322_d
    END SERIES_DIFFERENCE
```

Figure 32. Example of a SERIES_DIFFERENCE block.

The outcome of processing a SERIES_DIFFERENCE block is another time series, whose terms are the differences between successive terms of an existing time series. The new time series has one less term than the original time series; the date and time attributed to the first value of the new series is that of the second value present in the original series. For example, if a series spans the period 1/1/1990 to 2/1/1990, the value of the first term of the new series represents the January 2 value minus the January 1 value; the date associated with this difference would be, in this case, January 2, 1990.

SERIES_DISPLACE

The SERIES_DISPLACE block is used to migrate the terms of a series with respect to its time base, lagging or leading these terms as requested by the user. Keywords pertaining to the SERIES_DISPLACE block are listed in table 24. An example SERIES_DISPLACE block is given in figure 33. Keywords can be supplied in any order, except for the CONTEXT keyword(s), which must precede all others.

Table 24.   Keywords within a SERIES_DISPLACE block.

Keyword	Role	Specifications
CONTEXT	At least one CONTEXT keyword must be supplied; up to five are permitted. If one of the CONTEXT strings matches the CONTEXT string in the SETTINGS block, or if one of the CONTEXT strings is "all," the SERIES_DISPLACE block will be processed.	Any string without internal spaces of 20 or fewer characters. The CONTEXT keyword(s) must precede all other keywords.
SERIES_NAME	Mandatory. The name of the time series whose terms are to be displaced.	A name of 18 or fewer characters referencing a time series stored within TSPROC's memory.
NEW_SERIES_NAME	Mandatory. The name of the new time series formed by term displacement of an existing series.	Any character string without internal spaces up to 18 characters.
LAG_INCREMENT	Mandatory. The number of sample intervals by which terms of the original time series are lagged.	An integer.
FILL_VALUE	Mandatory. Values assigned to migrated terms at the beginning or end of the time series, where no other terms can take their place	A real number.

```
   START SERIES_DISPLACE
    CONTEXT all
    SERIES_NAME outflow
    NEW_SERIES_NAME outflow_1
    LAG_INCREMENT 1
    FILL_VALUE 0.00
   END SERIES_DISPLACE
```

Figure 33.   Example of a SERIES_DISPLACE block.

The SERIES_DISPLACE operation requires that the time series upon which the operation is carried out has a constant sample interval. If the sample interval is not constant throughout the time spanned by the time series, TSPROC displays an error message before ceasing execution.

A positive LAG_INCREMENT is used to delay terms in the time series. For example, if a LAG_INCREMENT of 1 is used, then each term within a time series is assigned to the time and date previously occupied by the term that follows it. If it is desired that terms in the series be shifted in the opposite direction instead, this can be accomplished by use of a negative LAG_INCREMENT.

When terms of a time series are shifted, terms at one end of the series "drop off the edge" (the time base of the series is not altered by the SERIES_DISPLACE operation). At the other end of the series, at least one term of the shifted series must be assigned a "dummy value," because end positions within the series become vacated by the shifting operation. The user must provide this "dummy value" by use of the FILL_VALUE keyword.

In undertaking sophisticated and powerful parameter-estimation procedures such as those described by Kuczera (1983), a combination of an original and a lagged "observed time series" must be compared with its model-generated counterpart. Residuals (differences between simulated and observed values) achieved by using combinations of original and lagged time series are often superior to those achieved by using only the original time series, because the former have drastically reduced inherent interterm correlation structure. The existence of such interterm correlation can lead to misleading estimates of parameter uncertainty.

A composite series, comprised of an original time series summed with various combinations of lagged time series, can be created using the SERIES_EQUATION block.

SERIES_EQUATION

The SERIES_EQUATION block is used to form a new time series based on an equation of arbitrary mathematical complexity involving one or several other time series. The only two conditions on time series that are cited in this equation are that:

1. all time series featured in the series equation must have samples at identical dates and times (this can be ensured by use of the REDUCE_TIME_SPAN and NEW_TIME_BASE blocks, if necessary); and

2. a series equation must feature at least one time series (to provide the time base of the resulting time series).

Keywords appearing in a SERIES_EQUATION block are listed table 25. An example of a SERIES_EQUATION block is shown in figure 34.

Table 25.   Keywords within a SERIES_EQUATION block.

Keyword	Role	Specifications
CONTEXT	At least one CONTEXT keyword must be supplied; up to five are permitted. If one of the CONTEXT strings matches the CONTEXT string in the SETTINGS block, or if one of the CONTEXT strings is "all," the SERIES_EQUATION block will be processed.	Any string without internal spaces of 20 or fewer characters. The CONTEXT keyword(s) must precede all other keywords.
NEW_SERIES_NAME	Mandatory. The name of the new time series formed through undertaking the calculations embodied in the series equation.	Any character string without internal spaces up to 18 characters.
EQUATION	Mandatory. The equation by which the terms of the new series are calculated.	See below.

```
START SERIES_EQUATION
 CONTEXT all
 NEW_SERIES_NAME new_series
 EQUATION log10(outflow * concentration)
END SERIES_EQUATION
```

Figure 34.   Example of a SERIES_EQUATION block.

The terms of the new time series are computed by implementing the equation on a term-by-term basis on each of the series cited in the equation. Thus, each term in the new series is calculated from the corresponding terms of the existing series.

The series equation can be of arbitrary complexity, involving any number of terms, and citing any number of existing time series, as long as the above mentioned time base-consistency rule is followed. In formulating the equation, the operators ^, /, *, -, and + have their usual meanings of "raised to the power of," "division," "multiplication," "subtraction," and "addition"; optionally the ** operator can be used in place of the ^ operator to signify raising to the power. Operations are carried out in the order indicated above; if in doubt, use parentheses to set precedance between operators. A summary of all allowable operations within a SERIES_EQUATION block is given in table 26.

Table 26. Allowable operations within a SERIES_EQUATION block.

[dd/mm/yyyy_hh:nn:ss,where dd is the number of digits for representing the day, mm represents the month, and yyyy or yy represents the year; hh:nn:ss, where hh is the number of digits representing the hour, nn represents minutes, and ss represents seconds]

Operator	Description
(x + y)	Grouping.
x ^ y	x raised to the power of y.
x ** y	x raised to the power of y.
x * y	Multiplication.
x / y	Division.
x - y	Subtraction.
x + y	Addition.
Trigonometric functions	**Description**
cos(x)	Cosine.
sin(x)	Sine.
tan(x)	Tangent.
acos(x)	Arc cosine.
asin(x)	Arc sine.
atan(x)	Arc tangent.
cosh(x)	Hyperbolic cosine.
sinh(x)	Hyperbolic sine.
tanh(x)	Hyperbolic tangent.
Miscellaneous functions	**Description**
abs(x)	Absolute value of x.
exp(x)	Exponential of x: e^x.
log(x)	Logarithm base e of x.
log10(x)	Logarithm base 10 of x.
@_days_start_year	Accepts no arguments, returns the days since start of year for each date represented in other time series in the equation.
@_days_"mm/dd/yyyy_hh:nn:ss"	Accepts no arguments, returns the days since "mm/dd/yyyy_hh:nn:ss" for each date represented in other time series in the equation.

An equation supplied in the SERIES_EQUATION block can include most of the commonly used mathematical functions: abs, acos, asin, atan, cos, cosh, exp, log, log10, sin, sinh, sqrt, tan, and tanh. Note the following standards for functions.

1.   The log function is to base e; to calculate logs to base 10, use the log10 function.

2.   The arguments to trigonometric functions must be supplied in radians.

3.   When some of these functions are used, their arguments must lie within the proper numerical range for that function. For example, if any of the terms of a series upon which a log operation is performed are zero or negative, a numerical error results. TSPROC detects this error and displays an error message before ceasing execution. When the / operator is used, a divide-by-zero condition must not encountered. If this occurs, TSPROC issues an error message before ceasing execution.

In addition to the above functions, TSPROC defines two other specialized functions for use in a series equation; these are the @_days_start_year and the @_days_"mm/dd/yyyy_hh:nn:ss" functions. The "@_" string indicates to the subroutine that parses this equation that the term represents neither a series, a number, nor one of the mathematical functions discussed above. Minutes in the SERIES_EQUATION block functions are represented by nn, rather than mm.

When the @_days_start_year term is encountered in a series equation, the days since the start of the year pertaining to the current series term are substituted for the string. Where a sample does not occur at midnight, fractional days are used in the calculation of the @_days_start_year function, whose outcome is a real number.

When the @_days_"mm/dd/yyyy_hh:nn:ss" term is encountered, TSPROC calculates the days (as a real number, fractional if necessary) since the indicated date and time. Note that the date and time strings must be collectively enclosed in quotes (") and must be separated by an underscore. In addition, the correct format for expressing the date (mm/dd/yyyy or dd/mm/yyyy) is determined by the DATE_FORMAT keyword in the SETTINGS block.

An equation may be as simple as the first of the equations as shown in figure 35. Although this equation is not terribly useful, applying it would result in a new series created by use of the unaltered terms of the existing series outflow.

```
outflow
log10(outflow) + 3.456 * sediment ^ 3.23
34.5 / (interflow + 3.432)
0.0 * series1 + @_days_start_year
3.495 + sin((@_days_start_year + 124.5)*6.284/365.25)
1.0/sqrt(@_days_"1/21/1978_12:00:00")
```

Figure 35.   Examples of valid EQUATION arguments.

In the fourth of the above equations, the time series named series1 is multiplied by zero. In this case, the series is included in the equation because each equation must cite at least one time series to set the time base of the resultant time series. In the fifth of the above equations, the argument of the sine function is multiplied by $\frac{2\pi}{365.25}$ to achieve periodicity of one year.

For those unfamiliar with programming, the equation $a/b*c$ is evaluated as $(a/b)*c$. To divide a by $b*c$, formulate the equation as: $a/(b*c)$ or $a/b/c$.

SERIES_STATISTICS

The SERIES_STATISTICS block is used to calculate simple statistics from the terms of a time series. Optionally, the terms of the series upon which statistical calculations are based can be limited to those within a specified date/time interval. Another option provided by the SERIES_STATISTICS block is for calculation of statistics on the basis of the log (to base 10) of the terms of the time series or on the terms of the series raised to an arbitrary power. If it is desired that statistics be calculated on the basis of more complex functions of the terms of a time series, this can be easily achieved by first calculating a new time series using the SERIES_EQUATION block, and then undertaking statistical calculations on the basis of this new time series.

Currently nine statistical measures can be calculated using the SERIES_STATISTICS block: the mean, median, standard deviation, sum, maximum, minimum, range, the minimum n-point mean, and the maximum n-point mean. If the user intends to use any statistics in a calibration exercise undertaken by PEST, then only those statistics that are actually involved in the parameter-estimation process should be calculated in a SERIES_STATISTICS block. This limits the output from the LIST_OUTPUT block to only those statistics.

TSPROC stores the outcomes of statistical calculations carried out by the SERIES_STATISTICS block in an S_TABLE. Like other TSPROC entities, each S_TABLE must be provided with a name so that it can be referenced by other TSPROC processing blocks. This name must be 18 or fewer characters and must not include a space character.

Keywords featured in the SERIES_STATISTICS block are listed in table 27. An example of a SERIES_STATISTICS block is given in figure 36.

Table 27.   Keywords in a SERIES_STATISTICS block.—Continued

[dd/mm/yyyy or mm/dd/yy, where dd is the number of digits for representing the day, mm represents the month, and yyyy or yy represents the year; hh:mm:ss, where hh is the number of digits representing the hour, mm represents minutes, and ss represents seconds]

Key word	Role	Specifications
CONTEXT	At least one CONTEXT keyword must be supplied; up to five are permitted. If one of the CONTEXT strings matches the CONTEXT string in the SETTINGS block, or if one of the CONTEXT strings is "all," the SERIES_STATISTICS block will be processed.	Any string without internal spaces of 20 or fewer characters. The CONTEXT keyword(s) must precede all other keywords.
SERIES_NAME	Mandatory. The name of the time series on which statistical calculations will be carried out.	A name of 18 or fewer characters referencing a time series stored within TSPROC's memory.
NEW_S_TABLE_NAME	Mandatory. The name of the new S_TABLE used to store the outcomes of statistical calculations.	Any character string without internal spaces up to 18 characters.
SUM	Optional. Requests calculation of the sum of the terms of the time series.	"yes" or "no" Default is "no."
MEAN	Optional. Requests calculation of the mean of the terms of the time series.	"yes" or "no" Default is "no."
MEDIAN	Optional. Requests calculation of the median of the terms of the time series.	"yes" or "no" Default is "no."
MINMEAN_n	Optional. Requests calculation of the minimum n-count mean of terms of the series.	"yes" or "no" Default is "no."
MAXMEAN_n	Optional. Requests calculation of the maximum n-count mean of terms of the series.	"yes" or "no" Default is "no."
STD_DEV	Optional. Requests calculation of the standard deviation of the terms of the time series.	"yes" or "no" Default is "no."
MAXIMUM	Optional. Requests calculation of the maximum of the terms of the time series.	"yes" or "no" Default is "no."
MINIMUM	Optional. Requests calculation of the minimum of the terms of the time series.	"yes" or "no" Default is "no."
RANGE	Optional. Requests calculation of the difference between the maximum and minimum of the terms of the time series.	"yes" or "no" Default is "no."

The use of the POWER and LOG keywords may have unintended consequences. It is illegal for both of these keywords to be present within the same SERIES_STATISTICS block. Furthermore, the use of these keywords may cause numerical errors. In particular, if LOG is set to "yes" and if any of the terms of the time series are zero or negative, then TSPROC displays an error message and ceases execution. In addition, if a POWER with an absolute value of less than 1 is supplied and if any of the terms of the time series are negative, or if the POWER is negative and any of the terms of the time series are zero, then TSPROC likewise ceases execution and displays an error message instead of attempting this impossible calculation.

The MINMEAN_n and MAXMEAN_n statistics require further explanation. As is apparent from the above example, the user must supply an appropriate value for n. For example, if MINMEAN_5 is set to "yes," TSPROC calculates the minimum value of the running mean of five consecutive values of the series, and this calculation takes place over the length of the series or between the user-provided beginning and end dates. If both the MINMEAN_n and MAXMEAN_n keywords are supplied in the same SERIES_STATISTICS block, then n must be the same for both of these keywords. In addition, the LOG and POWER keywords must not be supplied in the same block as the MINMEAN_n and MAXMEAN_n keywords; if this is a problem, use the SERIES_EQUATION block to transform the series prior to use of the SERIES_STATISTICS block.

Table 27. Keywords in a SERIES_STATISTICS block.—Continued

[dd/mm/yyyy or mm/dd/yy, where dd is the number of digits for representing the day, mm represents the month, and yyyy or yy represents the year; hh:mm:ss, where hh is the number of digits representing the hour, mm represents minutes, and ss represents seconds]

Key word	Role	Specifications
LOG	Optional. Requests that statistics be calculated based on the logs (to base 10) of the terms of the time series.	"yes" or "no." Default is "no." The LOG keyword cannot be used if the POWER keyword is used.
POWER	Optional. Requests that statistics be calculated based on the terms of the time series raised to the nominated power.	A real number other than zero. Default is 1. The POWER keyword cannot be used if the LOG keyword is used.
DATE_1	Optional. Terms of the time series before TIME_1 on this date are not used in statistics calculations.	Either dd/mm/yyyy or mm/dd/yy, depending on the DATE_FORMAT setting in the SETTINGS block.
TIME_1	Optional. Terms of the time series before this time on DATE_1 are not used in statistical calculations.	hh:mm:ss .
DATE_2	Optional. Terms of the time series after TIME_2 on this date are not used in statistical calculations.	Either dd/mm/yyyy or mm/dd/yy, depending on the DATE_FORMAT setting in the SETTINGS block.
TIME_2	Optional. Terms of the time series after this time on DATE_2 are not used in statistical calculations.	hh:mm:ss .

```
START SERIES_STATISTICS
  CONTEXT all
  SERIES_NAME outflow
  NEW_S_TABLE_NAME outflow
  MEAN yes
  STANDARD_DEVIATION yes
  SUM yes
  MAXIMUM yes
  MINIMUM yes
  MINMEAN_5 yes
  MAXMEAN_5 yes
  POWER 0.5
  DATE_1 3/1/1976
  TIME_1 00:00:00
  DATE_2 3/3/1976
  TIME_2 00:00:00
END SERIES_STATISTICS
```

Figure 36. Example of a SERIES_DIFFERENCE block.

SETTINGS

The SETTINGS block differs from the other blocks in a TSPROC input file in that it must be the first block listed in this file and cannot be repeated. Two keywords, DATE and CONTEXT, must be used in a SETTINGS block. Table 28 lists the allowable keywords within a SETTINGS block. Figure 37 shows an example SETTINGS block.

Table 28. Keywords within a SETTINGS block.

[dd/mm/yyyy or mm/dd/yy, where dd is the number of digits for representing the day, mm represents the month, and yyyy or yy represents the years]

Keyword	Role	Specifications
DATE_FORMAT	Mandatory. Determines the format with which dates are represented in TSPROC input and output files.	Either dd/mm/yyyy or mm/dd/yyyy.
CONTEXT	Mandatory. Sets the context for the current TSPROC run, thus determining which blocks in the TSPROC input file are processed.	Any character string without internal spaces of 20 or fewer characters.

```
     START SETTINGS
      DATE FORMAT mm/dd/yyyy
      CONTEXT pest_input
     END SETTINGS
```

Figure 37.   Example of a SETTINGS block.

The DATE_FORMAT setting allows TSPROC to adapt to the different methods for representing the date in different countries. Currently, TSPROC understands two date formats: mm/dd/yyyy and dd/mm/yyyy.

A SETTINGS block can contain only one CONTEXT keyword. Through the use of various context arguments, the user may activate or inactivate combinations of TSPROC blocks as needed. Every other block used in a TSPROC input file must contain a minimum of one CONTEXT keyword, and a maximum of five, followed by a character string of 20 characters or less and without internal spaces. If any of these character strings match the CONTEXT character string provided in the SETTINGS block, or if any of these strings is supplied as "all," then that block is processed.

TSPROC blocks may be activated or inactivated by choosing the argument given for the CONTEXT keyword. This can be particularly useful when using TSPROC with PEST. In preparing for a PEST run, a user can set up a complex TSPROC input file that processes measured and model-generated time series, and then generates a PEST input dataset in which the terms of the processed measured time series act as "calibration targets" to which the terms of the processed model-generated time series are matched. If CONTEXT settings in the various TSPROC processing blocks are carefully selected, then the same TSPROC input file can be used by TSPROC in its capacity as a model post-processor simply by altering the run CONTEXT in the SETTINGS block. Note that the context contained in the SETTINGS block may be overridden at the command line, as discussed in the section "The SETTINGS Block."

USGS_HYSEP

The USGS_HYSEP block can be used to separate baseflow from total flow for specific periods within a time series. The user may select one of three different baseflow separation methods for application within this block.

TSPROC stores the outcomes of baseflow separation performed by the USGS_HYSEP block in a new time series. Like other TSPROC entities, the new time series must be provided with a name so that it can be referenced by other TSPROC processing blocks. This name must be 18 or fewer characters and must not include a space character.

Keywords featured in the USGS_HYSEP block are listed in table 29. An example of a USGS_HYSEP block is given in figure 38.

Table 29. Keywords within a USGS_HYSEP block.

[dd/mm/yyyy or mm/dd/yy, where dd is the number of digits for representing the day, mm represents the month, and yyyy or yy represents the year; hh:mm:ss, where hh is the number of digits representing the hour, mm represents minutes, and ss represents seconds]

Keyword	Role	Specifications
CONTEXT	At least one CONTEXT keyword must be supplied; up to five are permitted. If one of the CONTEXT strings matches the CONTEXT string in the SETTINGS block, or if one of the CONTEXT strings is "all." the PERIOD_STATISTICS block will be processed.	Any string without internal spaces of 20 pr fewer characters . The CONTEXT keyword(s) must precede all other keywords.
SERIES_NAME	Mandatory. The name of the time series on which baseflow separations will be carried out.	A name of 18 or fewer characters referencing a time series stored within TSPROC's memory.
NEW_SERIES_NAME	Mandatory. The name of the new time series used to store the outcome of baseflow separation.	Any character string without internal spaces up to 18 characters.
HYSEP_TYPE	Mandatory. Keyword indicating the method to be used for baseflow separation.	"fixed_interval," "sliding_interval," or "local_minimum"
TIME_INTERVAL	Mandatory. Time interval to be used for baseflow separation, in days.	A non-negative odd number.
DATE_1	Optional. Terms of the time series before TIME_1 on this date are not used in peak extraction.	Either dd/mm/yyyy or mm/dd/yy, depending on the DATE_FORMAT setting in the SETTINGS block.
TIME_1	Optional. Terms of the time series before this time on DATE_1 are not used in peak extraction.	hh:mm:ss .
DATE_2	Optional. Terms of the time series after TIME_2 on this date are not used in peak extraction.	Either dd/mm/yyyy or mm/dd/yy, depending on the DATE_FORMAT setting in the SETTINGS block.
TIME_2	Optional. Terms of the time series after this time on DATE_2 are not used in peak extraction.	hh:mm:ss .

```
    START USGS_HYSEP
    CONTEXT obs_prep
    SERIES_NAME o_001_d
    NEW SERIES NAME o_001_bf
    HYSEP_TYPE local_minimum
    TIME_INTERVAL 3
    DATE_1 10/05/1991
    TIME_1 0:00:00
    DATE_2 09/30/1999
    TIME_2 0:00:00
    END USGS_HYSEP
```

Figure 38. Example of a USGS_HYSEP block.

The techniques used for baseflow separation are described in Sloto and Crouse (1996). Each technique uses a time interval to perform the separation, which is usually based on the following formula for time of concentration (time interval following cessation of rainfall where surface runoff occurs):

$$t_c = A^{0.2}$$

(11)

where t_c is time of concentration in days, and A is drainage area in mi^2. In general, the interval used should be twice the width of the time of concentration, rounded to the nearest odd integer, although any odd integer can be entered for the TIME_INTERVAL keyword.

V_TABLE_TO_SERIES

The V_TABLE_TO_SERIES block copies information stored in a V_TABLE to a new time series. Information stored in time-series format has access to more processing functionality than that available for V_TABLEs, including calculation of comparison statistics with other series, digital filtering, time interpolation, and other operations.

Keywords associated with the V_TABLE_TO_SERIES block are listed in table 30. An example of a V_TABLE_TO_SERIES block is given in figure 39.

Table 30. Keywords in a V_TABLE_TO_SERIES block.

Keyword	Role	Specifications
CONTEXT	At least one CONTEXT keyword must be supplied; up to five are permitted. If one of the CONTEXT strings matches the CONTEXT string in the SETTINGS block, or if one of the CONTEXT strings is "all," the V_TABLE_TO_SERIES block will be processed.	Any string without internal spaces of 20 or fewer characters. The CONTEXT keyword(s) must precede all other keywords.
NEW_SERIES_NAME	Mandatory. The name of the new time series formed through copying entries from a V_TABLE.	Any character string without internal spaces up to 18 characters.
V_TABLE_NAME	Mandatory. The name of a V_TABLE from which entries are to be copied to the new time series.	A name of 18 or fewer characters a V_TABLE stored within TSPROC's memory.
TIME_ABSCISSA	Mandatory. Informs TSPROC whether the date and time corresponding to each new time series entry pertains to the beginning, middle, or end of the corresponding V_TABLE interval.	"start," "center," or "end" .

```
    START V_TABLE_TO_SERIES
     CONTEXT all
     V_TABLE_NAME volume
     NEW_SERIES_NAME ssvol
     TIME_ABSCISSA end
    END V_TABLE_TO_SERIES
```

Figure 39. Example of a SETTINGS block.

Although two dates and times are associated with every term of a V_TABLE (corresponding to the interval over which volume is accumulated), only one date and time is associated with every time-series entry. In the process of converting from a table to a series, the user must inform TSPROC how time-series dates and times are calculated from V_TABLE dates and times. Three options are available:

1. Time-series dates and times can correspond to the beginnings of respective volume accumulation intervals of the V_TABLE from which they are derived;

2. Time-series dates and times can correspond to the ends of respective volume accumulation intervals of the V_TABLE from which they are derived;

3. Time-series dates and times can correspond to the centers of respective volume accumulation intervals of the V_TABLE from which they are derived.

Select the appropriate option by providing the character string "start," "end," or "center" (or "centre") with the TIME_ABSCISSA keyword of a V_TABLE_TO_SERIES block.

VOLUME_CALCULATION

The VOLUME_CALCULATION block instructs TSPROC to integrate a time series with respect to time over the time span bracketed by two dates and times. Although the most obvious application of this functionality is in volume calculation, it can also be used for mass calculation if the integration is carried out on a time series that represents the mass flux of some constituent. A mass flux time series can be calculated from time series representing concentration and flow by use of the SERIES_EQUATION block.

Integration can be carried out over one or multiple time spans defined in a dates file; the format of a dates file is shown in figure 40. Dates and times are supplied in a dates file rather than as part of the VOLUME_CALCULATION block, because a large number of volumes or constituent masses may often be used in the calibration process. Integration may take place over regularly spaced (for example monthly) time intervals or over a number of discrete, significant events.

```
03/12/1976 11:23:53   04/03/1976 03:00:00
04/30/1976 12:43:00   09/02/1976 23:59:59
04/30/1976 12:43:00   04/30/1976 23:59:59
```

Figure 40.   Example of a dates file.

A dates file can be of any length. Each line must contain four entries: the date and time defining the beginning of the integration interval, and the date and time defining the end of the interval. The date format must be dd/mm/yyyy or mm/dd/yyyy, and it must be consistent with the DATE_FORMAT setting in the TSPROC SETTINGS block.

The outcomes of TSPROC's volume calculations are stored in a V_TABLE. Like other TSPROC entities, each V_TABLE must be given a name, supplied through the NEW_V_TABLE_NAME keyword. This and other keywords associated with a VOLUME_CALCULATION block are listed in table 31. An example of a VOLUME_CALCULATION block is given in figure 41.

Table 31.   Keywords in a VOLUME_CALCULATION block.

[min, minutes; sec, seconds]

Keyword	Role	Specifications
CONTEXT	At least one CONTEXT keyword must be supplied; up to five are permitted. If one of the CONTEXT strings matches the CONTEXT string in the SETTINGS block, or if one of the CONTEXT strings is "all," the VOLUME_CALCULATION block will be processed.	Any string without internal spaces of 20 or fewer characters. The CONTEXT keyword(s) must precede all other keywords.
SERIES_NAME	Mandatory. The name of the time series on which time integration will be carried out.	A name of 18 or fewer characters referencing a time series stored within TSPROC's memory.
NEW_V_TABLE_NAME	Mandatory. The name of a new V_TABLE used to store the outcomes of time-integration carried out by TSPROC.	Any character string without internal spaces up to 18 characters.
DATE_FILE	Mandatory. The name of the dates file containing the time spans over which time series integration will take place.	Any filename up to 120 characters. Use quotes if the filename contains spaces.
FLOW_TIME_UNITS	Mandatory. The time units of flow employed by the time series.	"year," "month," "day," "hour," "min," or "sec".
FACTOR	Optional. Factor by which integrated volumes or masses are multiplied before storage	A real number. Default is 1.0.

```
START VOLUME_CALCULATION
 CONTEXT all
 SERIES_NAME outflow
 NEW_V_TABLE_NAME volout
 FLOW_TIME_UNITS days
 DATE_FILE "volume dates.dat"
 FACTOR 3.4953
END VOLUME_CALCULATION
```

Figure 41. Example of a VOLUME_CALCULATION block.

Two VOLUME_CALCULATION keywords require further explanation. By use of the required FLOW_TIME_UNITS keyword, the user supplies the time units employed by the flow time series. For example, if flow is recorded in cubic feet per second, then FLOW_TIME_UNITS should be provided as "sec." By use of the optional FACTOR keyword, the user supplies a multiplier that TSPROC applies to each integrated volume or mass that it calculates. The predominant use of this multiplier is in units conversion. For example, if the volume in cubic feet calculated in the above example is to be stored in units of acre feet, gallons, megalitres, or some other volumetric unit, then the appropriate conversion factor should be supplied.

WRITE_PEST_FILES

The WRITE_PEST_FILES block instructs TSPROC to generate PEST input files for a parameter-estimation run. The CONTEXT keyword in the SETTINGS block should be used to activate the WRITE_PEST_FILES block during PEST control file generation, and it is also used to deactivate this block when TSPROC is used as part of a composite model.

It is beyond the scope of this publication to discuss parameter estimation using PEST software on more than a superficial basis; readers should consult PEST documentation (Doherty, 2010a, b) for more detail on the use and application of PEST.

Position within a TSPROC Input File

If present, a WRITE_PEST_FILES block must immediately follow a LIST_OUTPUT block in a TSPROC input file. In writing the PEST input dataset, TSPROC assumes that the LIST_OUTPUT block that immediately precedes the WRITE_PEST_ FILES block is exactly the same as that which it uses to generate model output files when run as a model post-processor during a calibration run. The time series and tables cited in the LIST_OUTPUT block are generated by the model under calibration. For each of these model-generated entities, a corresponding observation entity must be supplied. Like the model entities to which they are matched, the observation entities must have been generated, or simply imported, during the current TSPROC run.

Model and Observation Entities

Model-generated and observed time series and tables must be comparable; they must contain the same number of values and cover identical date and time ranges. Should TSPROC detect any inconsistencies in such paired entities, it displays an appropriate error message before ceasing execution.

Time interpolation of model-generated series to the times and dates of their corresponding observation time series is advisable before any further analyses are performed, especially if observations are intermittent and irregular. By doing this, any bias or miscalculation of the quantities stored within the various TSPROC entities is "cancelled out" in the calibration process, because both the model and observation quantities are subject to exactly the same error caused by limitations in the time base on which they were calculated.

Keywords

Table 32 describes the keywords associated with a WRITE_PEST_FILES block.

An example of a WRITE_PEST_FILES block is given in figure 42. Note that the WRITE_PEST_FILES block may include an almost unlimited number of paired comparisons between model-generated and observed series and tables.

Table 32. Keywords in a WRITE_PEST_FILES block.

Keyword	Role	Specifications
CONTEXT	At least one CONTEXT keyword must be supplied; up to five are permitted. If one of the CONTEXT strings matches the CONTEXT string in the SETTINGS block, or if one of the CONTEXT strings is "all," the WRITE_PEST_FILES block will be processed.	Any string without internal spaces of 20 or fewer characters. The CONTEXT keyword(s) must precede all other keywords.
TEMPLATE_FILE	Mandatory. The name of a PEST template file. Use as many TEMPLATE_FILE entries as there are template files involved in the parameter estimation process.	Any filename up to 120 characters. Use quotes if the filename contains spaces.
MODEL_INPUT_FILE	Optional. If present, this keyword must immediately follow a TEMPLATE_FILE keyword.	Any filename up to 120 characters. Use quotes if the filename contains spaces.
PARAMETER_DATA_FILE	Optional. The name of a file containing data normally found in the "parameter data" section of a PEST control file.	Any filename up to 120 characters. Use quotes if the filename contains spaces.
PARAMETER_GROUP_FILE	Optional. The name of a file containing data normally found in the "parameter groups" section of a PEST control file.	Any filename up to 120 characters. Use quotes if the filename contains spaces.
OBSERVATION_SERIES_NAME	Optional. The name of a time series containing measurement data. Must be followed by a MODEL_SERIES_NAME keyword.	A name of 18 or fewer characters referencing a time series stored within TSPROC's memory.
MODEL_SERIES_NAME	Mandatory for every OBSERVATION_SERIES_NAME keyword. Model-generated time series corresponding to an observation time series. Must follow an OBSERVATION_SERIES_NAME keyword.	A name of 18 or fewer characters referencing a time series stored within TSPROC's memory.
SERIES_WEIGHTS_EQUATION	Mandatory for every OBSERVATION_SERIES_NAME keyword. Equation by which observation weights are calculated. Must follow a MODEL_SERIES_NAME keyword.	An equation, optionally enclosed in quotes.
SERIES_WEIGHTS_MIN_MAX	Optional. The minimum and maximum weights for observations pertaining to the previous OBSERVATION_SERIES_NAME keyword. If present, must immediately follow a SERIES_WEIGHTS_EQUATION keyword.	Two real nonnegative numbers separated by a space. First the minimum weight, then the maximum weight.
OBSERVATION_S_TABLE_NAME	Optional. The name of an S_TABLE containing processed measurement data. Must be followed by a MODEL_S_TABLE_NAME keyword.	A name of 18 or fewer characters referencing an S_TABLE stored within TSPROC's memory.
MODEL_S_TABLE_NAME	Mandatory for every OBSERVATION_S_TABLE_NAME keyword. Model-generated S_TABLE corresponding to an observation S_TABLE. Must follow an OBSERVATION_S_TABLE_NAME keyword.	A name of 18 or fewer characters referencing an S_TABLE stored within TSPROC's memory.
S_TABLE_WEIGHTS_EQUATION	Mandatory for every OBSERVATION_S_TABLE_NAME keyword. Equation by which observation weights are calculated. Must follow a MODEL_S_TABLE_NAME keyword.	An equation, optionally enclosed in quotes.
S_TABLE_WEIGHTS_MIN_MAX	Optional. The minimum and maximum weights for observations pertaining to the previous OBSERVATION_S_TABLE_NAME keyword. If present, must immediately follow an S_TABLE_WEIGHTS_EQUATION keyword.	Two real nonnegative numbers separated by a space. First the minimum weight, then the maximum weight.
OBSERVATION_V_TABLE_NAME	Optional. The name of a V_TABLE containing processed measurement data. Must be followed by a MODEL_V_TABLE_NAME keyword.	A name of 18 or fewer characters referencing a V_TABLE stored within TSPROC's memory.

Table 32.   Keywords in a WRITE_PEST_FILES block.—Continued

Keyword	Role	Specifications
MODEL_V_TABLE_NAME	Mandatory for every OBSERVATION_V_TABLE_NAME keyword. Model-generated V_TABLE corresponding to an observation V_TABLE. Must follow an OBSERVATION_V_TABLE_NAME keyword.	A name of 18 or fewer characters referencing a V_TABLE stored within TSPROC's memory.
V_TABLE_WEIGHTS_EQUATION	Mandatory for every OBSERVATION_V_TABLE_NAME keyword. Equation by which observation weights are calculated. Must follow a MODEL_V_TABLE_NAME keyword.	An equation, optionally enclosed in quotes.
V_TABLE_WEIGHTS_MIN_MAX	Optional. The minimum and maximum weights for observations pertaining to the previous OBSERVATION_V_TABLE_NAME keyword. If present, must immediately follow a V_TABLE_WEIGHTS_EQUATION keyword.	Two real nonnegative numbers separated by a space. First the minimum weight, then the maximum weight.
OBSERVATION_E_TABLE_NAME	Optional. The name of an E_TABLE containing processed measurement data. Must be followed by a MODEL_E_TABLE_NAME keyword.	A name of 18 or fewer characters referencing an E_TABLE stored within TSPROC's memory.
MODEL_E_TABLE_NAME	Mandatory for every OBSERVATION_E_TABLE_NAME keyword. Model-generated E_TABLE corresponding to an observation E_TABLE. Must follow an OBSERVATION_E_TABLE_NAME keyword.	A name of 18 or fewer characters referencing an E_TABLE stored within TSPROC's memory.
E_TABLE_WEIGHTS_EQUATION	Mandatory for every OBSERVATION_E_TABLE_NAME keyword. Equation by which observation weights are calculated. Must follow a MODEL_E_TABLE_NAME keyword.	An equation, optionally enclosed in quotes.
E_TABLE_WEIGHTS_MIN_MAX	Optional. The minimum and maximum weights for observations pertaining to the previous OBSERVATION_E_TABLE_NAME keyword. If present, must immediately follow an E_TABLE_WEIGHTS_EQUATION keyword.	Two real nonnegative numbers separated by a space. First the minimum weight, then the maximum weight.
OBSERVATION_G_TABLE_NAME	Optional. The name of an G_TABLE containing processed measurement data. Must be followed by a MODEL_G_TABLE_NAME keyword.	A name of 18 or fewer characters referencing a G_TABLE stored within TSPROC's memory.
MODEL_G_TABLE_NAME	Mandatory for every OBSERVATION_G_TABLE_NAME keyword. Model-generated G_TABLE corresponding to an observation G_TABLE. Must follow an OBSERVATION_G_TABLE_NAME keyword.	A name of 18 or fewer characters referencing a G_TABLE stored within TSPROC's memory.
G_TABLE_WEIGHTS_EQUATION	Mandatory for every OBSERVATION_G_TABLE_NAME keyword. Equation by which observation weights are calculated. Must follow a MODEL_G_TABLE_NAME keyword.	An equation, optionally enclosed in quotes.
G_TABLE_WEIGHTS_MIN_MAX	Optional. The minimum and maximum weights for observations pertaining to the previous OBSERVATION_G_TABLE_NAME keyword. If present, must immediately follow an G_TABLE_WEIGHTS_EQUATION keyword.	Two real nonnegative numbers separated by a space. First the minimum weight, then the maximum weight.
NEW_PEST_CONTROL_FILE	Mandatory. The name of the PEST control file to be written by TSPROC.	Any filename up to 120 characters. Use quotes if the filename contains spaces.

Table 32. Keywords in a WRITE_PEST_FILES block.—Continued

Keyword	Role	Specifications
AUTOMATIC_USER_INTERVEN-TION	Optional. Determines the "automatic user intervention" setting in the PEST control file written by TSPROC.	"yes" or "no".
TRUNCATED_SVD	Optional. Determines whether truncated singular value decomposition will be employed as a stabilization device in the PEST control file written by TSPROC.	A real number greater than zero—the PEST EIGTHRESH variable. Normally about 2.0E-7.
NEW_INSTRUCTION_FILE	Mandatory. The name of the instruction file to be written by TSPROC.	Any filename up to 120 characters. Use quotes if the filename contains spaces.
MODEL_COMMAND_LINE	Optional. The model command line to be recorded in the "model command line" section of the PEST control file.	A command line that satisfies the requirements of the operating system, in the present case the name of a batch or script file.

```
    START WRITE_PEST_FILES
    CONTEXT pest_input
    NEW_PEST_CONTROL_FILE case.pst
    AUTOMATIC_USER_INTERVENTION yes
    TEMPLATE_FILE catchment.tpl
    MODEL_INPUT_FILE catchment.uci
    NEW_INSTRUCTION_FILE observation.ins

    OBSERVATION_SERIES_NAME flow_obs
    MODEL_SERIES_NAME i_flow_mod
    SERIES_WEIGHTS_EQUATION 1.0/@_abs_value
    SERIES_WEIGHTS_MIN_MAX 1.0 1000.0

    OBSERVATION_V_TABLE_NAME vol_obs
    MODEL_V_TABLE_NAME vol_mod
    V_TABLE_WEIGHTS_EQUATION 5.0

    OBSERVATION_S_TABLE_NAME stat_obs
    MODEL_S_TABLE_NAME stat_mod
    S_TABLE_WEIGHTS_EQUATION 1.0/@_abs_value

    OBSERVATION_E_TABLE_NAME time_obs
    MODEL_E_TABLE_NAME time_mod
    E_TABLE_WEIGHTS_EQUATION log(2.0/@_abs_value) + 2.0
    E_TABLE_WEIGHTS_MIN_MAX 0 1000

    PARAMETER_DATA_FILE param.dat
    PARAMETER_GROUP_FILE pargroup.dat
    MODEL_COMMAND_LINE model.bat
    END WRITE_PEST_FILES
```

Figure 42. Example of a WRITE_PEST_FILES block.

Tasks Undertaken by TSPROC in Generating a PEST Input Dataset

In processing the entries contained within a WRITE_PEST_FILES block, TSPROC performs the following tasks.

1. TSPROC reads all template files cited in the WRITE_PEST_FILES block, accumulating the names of all parameters cited in those files.

2. If a PARAMETER_DATA_FILE keyword is present within the WRITE_PEST_FILES block, TSPROC reads that file and stores the data for later use.

3. If a PARAMETER_GROUP_FILE keyword is present within the WRITE_PEST_FILES block, TSPROC reads that file and stores the data for later use.

4. TSPROC checks that all model series and tables cited in the WRITE_PEST_FILES block are also cited in the LIST_ OUTPUT block that should immediately precede it in the TSPROC input file.

5. TSPROC checks that each observation entity that is matched to a model entity is comparable; in other words, it checks to ensure that table comparisons are being made between identical table types and that series contain identical numbers of values and cover identical time and date ranges.

6. TSPROC then generates names for all observations featured in the parameter-estimation process.

7. TSPROC writes an instruction file by which the model-generated data written by the previous LIST_OUTPUT block can be read by PEST.

8. TSPROC then writes the "control data," "parameter group," and "parameter data" sections of the new PEST control file. Included in this file are all parameters referenced in the template files cited in the WRITE_PEST_FILES block. Information contained within the parameter data and parameter group files is included in the pertinent sections of the PEST control file where appropriate. Default values are used for all other PEST variables.

9. TSPROC then writes the "observation group" and "observation data" sections of the new PEST control file. Observation weights are calculated according to formulae supplied through various WEIGHTS_EQUATION keywords.

10. TSPROC writes the "model command line" and "model input/output" sections of the new PEST control file.

These tasks are now discussed in greater detail.

Parameter Data and Parameter Group

TSPROC accumulates the names of the parameters that it must include in the PEST control file by reading all template files cited in the WRITE_PEST_FILES block. Any number of TEMPLATE_FILE keywords can be included in a WRITE_PEST_ FILES block ; each TEMPLATE_FILE keyword should be followed by a MODEL_INPUT_FILE keyword. PEST links the model input file to the preceding template file when writing the "* model input/output" section of the PEST control file. If a MODEL_INPUT_FILE keyword is not associated with a particular TEMPLATE_FILE keyword, PEST supplies a default model input filename to correspond to the template file; this filename must be altered to the correct filename in the PEST control file before running PEST.

In writing a PEST control file, TSPROC must supply each parameter with an initial value, an upper and lower bound, and all of the other information contained within the "parameter data" section of a PEST control file. It must also assign each parameter to a parameter group. Recall that variables that govern the calculation of derivatives are assigned to parameter groups rather than to individual parameters. For some parameter types, the values assigned to these derivative-calculation variables can be crucial to the success of the parameter-estimation process.

If desired, default TSPROC parameter data can be overridden by supplying the values for parameter variables and parameter group variables through a "parameter data file" and a "parameter group file," respectively. The names of these files are supplied following optional keywords of the same name in the WRITE_PEST_FILES block.

If no PARAMETER_DATA_FILE keyword is present within a WRITE_PEST_FILES block, PEST assigns default values to all parameter variables. It assigns each parameter to a group of its own and supplies default values to the derivatives-calculation variables pertaining to each such group. The user should carefully inspect all of these variables, altering them as necessary to suit the calibration problem at hand.

An example of a parameter data file is given in figure 43.

```
rol fixed     factor  0.5  .1   10   ro  1.0  0.0
ro2 log       factor  5.0  .1   10   ro  1.0  0.0
ro3 tied ro1 factor  0.5  .1   10   ro  1.0  0.0
h1  none      factor  2.0  .05 100   h  1.0  0.0
h2  none      factor  5.0  .05 100   h  1.0  0.0
```

Figure 43. Example of a parameter data file.

A parameter data file is nearly identical to the "* parameter data" section of a PEST control file, containing the same variables in the same order. Differences between the parameter data file formats used by TSPROC and PEST include:

1. A value need not be supplied for the DERCOM variable, which is the command line number for derivatives calculation and the 10th variable on each line of the "parameter data" section of a PEST control file. TSPROC always provides a default value of 1 for this variable when it writes a PEST control file.

2. Not all parameters cited in template files need to be cited in a parameter data file. TSPROC provides default data for parameters that are absent from the latter file.

3. If a parameter is tied to another parameter, the name of the parent parameter must be attached to the "tied" string following an underscore, as illustrated in figure 43.

4. If a parameter is assigned to a particular parameter group, and if a parameter group file is not cited in the WRITE_PEST_FILES block, or if the name of the group is not included in a cited parameter group file, then TSPROC supplies default values for variables governing derivatives calculation for that group when it writes the PEST control file.

The contents of a parameter group file emulate those of the "parameter groups" section of a PEST control file. An example of a parameter group file is given in figure 44.

```
ro relative 0.01 0.00  switch 1.5 parabolic
h  relative 0.01 1.0e-4 switch 2.0 parabolic
```

Figure 44. Example of a parameter group file.

Time Series Observations

For every time series involved in the parameter-estimation process, at least three, and up to four, keywords must be supplied in the WRITE_PEST_FILES block. These keywords must be provided in the order presented in table 32.

The time series associated with the OBSERVATION_SERIES_NAME keyword should contain observation data. TSPROC writes the terms of this series to the PEST control file. The goal of the parameter-estimation process is to minimize the discrepancies between these terms and those of a corresponding model-generated time series. The model-generated time series is produced by TSPROC in its role as a model post-processor; as mentioned above, when TSPROC acts in this latter role, CONTEXT settings should be such that a PEST input dataset is NOT generated, and any unnecessary processing of observation data is avoided.

The name of the model time series that forms the model-generated counterpart to the observation time series must be supplied with the MODEL_SERIES_NAME keyword directly following the OBSERVATION_SERIES_NAME keyword. It is important that this same series is featured in the LIST_OUTPUT block immediately preceding the WRITE_PEST_FILES block. This LIST_OUTPUT block, and all calculations and data importations giving rise to the time series and tables cited in that block, must be retained when TSPROC is run as a model post-processor during the parameter-estimation process.

When writing a PEST input file, TSPROC assigns all observations comprised of the terms of an observation time series to a single observation group. This group is given the same name as the model time series to which the observation time series corresponds. Individual observation names are generated by affixing the string "n..n" to a contraction of the group name, where "n. n" is the term number of the time series. If for some reason this process does not result in unique observation names, which can occur under some circumstances if time series names are too similar, then TSPROC displays an error message and ceases execution.

When writing the "observation data" section of a PEST control file, TSPROC must assign a weight to each observation. Observation weights are calculated by TSPROC on the basis of the equation supplied by the user with the WEIGHTS_EQUATION keyword. The format of the weights equation is the same as that described in the SERIES_EQUATION block, except for two important differences.

1. If a series name is cited in a weights equation, that series must have the same time base (same number of terms and same date/time pertaining to each term) as the observation time series for which weights are being calculated. In implementing the equation for weights calculation, series are matched on a term-by-term basis.

2. An extra TSPROC-specific function, the @_abs_val function , is provided for use in a weights equation, but it is not available for use in a series equation. This function returns the absolute values of the terms of the observation time series for which a weight is currently being calculated.

Some example weights equations are given in figure 45.

```
wt_series
1.0/sqrt(@_abs_val)
4.0
1.0 + 0.5 * sin((@_days_start_year + 124.5)*6.284/365.25)
sqrt(@_days_"1/1/1989_00:00:00")
```

Figure 45. Example of valid weights equations.

In the first of the above equations, weights are simply set to the values of an existing time series. In the second of the above equations, observation weights are calculated as the inverse of the square root of the absolute value of each observation. In the third example, a uniform weight of 4.0 is assigned to all observations comprising the observation time series, while in the fourth example, weights show a seasonal dependance, being a function of time of year (note the factor of $\frac{2\pi}{365.25}$ in the argument to the sine function). Recall that the argument to the sin, cos, and tan functions must be supplied in radians; 2π radians is the same as 360 degrees. In the fifth of the above equations, weights increase as the square root of the number of days that have elapsed since the first moment of 1989.

If a negative observation weight is calculated , TSPROC raises the weight to zero. However, the user has the option of supplying upper and lower bounds to the weights through a SERIES_WEIGHTS_MIN_MAX keyword; if a user requests a minimum weight of less than 0.0, TSPROC overrides this with a minimum weight of zero.

S_TABLE Observations

The mechanism by which S_TABLE observations are included in a calibration dataset is very similar to that for including series observations in this dataset. The name of an observation S_TABLE must be provided through the OBSERVATION_S_TABLE keyword. This keyword must be immediately followed by a MODEL_S_TABLE keyword that provides the name of a corresponding model S_TABLE. This S_TABLE must contain the same statistics as those contained within the observation S_TABLE (statistics for inclusion in an S_TABLE are requested through the SERIES_STATISTICS block). This same S_TABLE must also be featured in the LIST_OUTPUT block immediately preceding the WRITE_PEST_FILES block.

TSPROC assigns all observations pertaining to a particular S_TABLE to a single observation group whose name is the same as that of the model S_TABLE. Individual members of the S_TABLE are provided with observation names by contracting the name of the observation group and appending a shortened form of the name of the statistic that each represents.

Weights for S_TABLE observations are generated by use of a weights equation. Unlike the weights equation used in the generation of weights for time series observations, the weights equation used for the generation of S_TABLE observation weights cannot cite a series name, nor can it use the @_days_start_year or @_days_"mm/dd/yyyy_hh:nn:ss" functions. However, it can use the @_abs_val function; in this case, the value refers to the particular statistical entity contained in the S_TABLE to which the weight is assigned.

V_TABLE Observations

V_TABLE observations are included in a calibration dataset in the same way that S_TABLE observations are included. The only difference is that individual observations are named by affixing a number (rather than a contracted form of the name of a statistical measure) to a contracted form of the observation group name. V_TABLE observations are named after the model V_TABLE to which the observation V_TABLE is matched in the WRITE_PEST_FILES block.

G_TABLE and E_TABLE Observations

Inclusion of G_TABLE or E_TABLE observations in the calibration process follows the same procedure as that used for inclusion of V_TABLE observations.

C_TABLE Observations

Data contained within C_TABLEs cannot be included in the model-calibration process. If the name of a C_TABLE is cited in a WRITE_PEST_FILES block, TSPROC displays an error message before ceasing execution.

The PEST Control File

The PEST control file written by TSPROC will likely need to be edited by the user to ensure that it is compatible with the optimization problem at hand. In the PEST control file written by TSPROC, PEST is asked to run in parameter-estimation mode. Default values are provided for all PEST control variables; these values are suitable for most occasions. If you would like PEST to run in regularization mode, you must also add a set of regularization observations and/or prior information equations to the TSPROC-generated PEST control file.

If a MODEL_COMMAND_LINE keyword is provided in a WRITE_PEST_FILES block, the user-supplied command line argument is transferred to the "model command line" section of the PEST control file written by TSPROC. Otherwise a default command line is supplied; it probably will need to be altered by the user before running PEST. Note that the model command line is the name of the composite model: the name of the batch or script file that runs the model or models along with PAR2PAR, TSPROC and any other glue code. Commands cited in this file will include the name of a model executable, as well as the command to run TSPROC.

The AUTOMATIC_USER_INTERVENTION keyword is used to set the DOAUI variable in the "control data" section of the PEST control file written by TSPROC. If this is set to "yes," then DOAUI is set to "aui"; PEST will implement automatic user intervention as necessary when implementing the inversion process. If AUTOMATIC_USER_INTERVENTION is set to "no," or if it is omitted, then DOAUI is set to "noaui." Regardless of the setting of this variable, TSPROC does not add an "automatic user intervention" section to the PEST control file that it writes. Thus, if automatic user intervention is implemented, it is done on the basis of default values for variables that control its implementation.

An alternative numerical stabilization device is truncated singular value decomposition ("truncated SVD"). This is invoked by supplying a TRUNCATED_SVD keyword, followed by the value of the truncation threshold (PEST variable EIGTHRESH), which is normally between 10^{-6} and 10^{-7}. TSPROC objects if both a TRUNCATED_SVD and an AUTOMATIC_USER_INTERVENTION keyword are supplied, because only one of these stabilization devices can be employed at the same time.

In addition, if truncated SVD is selected as a numerical stabilization device, then the initial lambda (RLAMBDA1 variable) is set to zero and the number of tested lambdas per iteration (NUMLAM variable) is set to 1 in the PEST control file written by TSPROC.

Although it may require some alterations before being used by PEST, a PEST control file written by TSPROC is complete enough to withstand the scrutiny of PESTCHEK. As is described in the PEST manual (Doherty, 2010a, b), PESTCHEK checks the PEST control file, whose name is provided on its command line, as well as all template and instruction files cited within the PEST control file.

Calibration by Using Patterns

The matching of raw or processed observation data to corresponding raw or processed model-generated data sometimes might not work as well as other strategies for at least some of the data types that may be included in the model-calibration process. For certain types of data , a better calibration strategy may be to attempt to match some relationship between flows and constituent observations (calculated on the basis of observations and model outputs, respectively), rather than the individual constituent concentrations themselves. For example, the calibration process may attempt to ensure that a regression relationship involving flows, constituent data, and, possibly, other factors, such as time of year, is respected by the model, even if the model is incapable of replicating individual constituent observations due to the erratic and noisy nature of these observations.

As an example of the application of this principal, consider that it is "known" that a certain regression relationship exists between flow and constituent concentrations. The coefficients in such a relationship may have been determined through using a model such as the USGS program ESTIMATOR, or they may even have been determined using PEST with TSPROC, with the SERIES_EQUATION block of TSPROC comprising the "model." As part of TSPROC's model post-processing duties, model-generated flows and constituent concentrations could be time interpolated to the dates and times when constituent observations were made. By use of the SERIES_EQUATION block, the difference between model-calculated concentrations and those "predicted" using the regression equation applied to model-generated flows could be evaluated. The closer that the difference between these two quantities is to zero, the closer does the "constituent pattern" generated by the model match the observed "constituent pattern." Other factors come into play here, such as the average and standard deviation of the constituent observations that, as discussed above, are also easily incorporated into the parameter-estimation process.

To incorporate "pattern matching" of this type into the parameter-estimation process, a time series expressing the difference between modeled constituent concentrations and those calculated from modeled flows using the "known" regression equation can be supplied as a model time series in the WRITE_PEST_FILES block (and the LIST_OUTPUT block preceding it). For consistency, dates and times for this time series should correspond only to constituent observation times. The corresponding observation time series would be one with an identical time base, but with all terms equal to zero. Weights assigned to these "observations" could be uniform; alternatively, they could be a function of the actual observed constituent concentrations, calculated by use of a SERIES_EQUATION block and supplied through the SERIES_WEIGHTS_EQUATION keyword.

References

Cocca, P.A., Doherty, J. and Kittle, J.L., Jr., 2004, Hydrologic calibration strategies for the HSPF watershed model: identifying effective objective functions for use with the Parameter Estimation (PEST) program, *in* World Water and Environmental Resources Congress and related symposia: World Water and Environmental Resources Congress, Philadelphia, Penn., 2003, p. 106.

Criss, R.E. and Winston, W.E., 2008, Do Nash values have value? Discussion and alternate proposals: Hydrological Processes, v. 22, no. 14, p. 2723–2725.

14, p. 2723–2725. Doherty, John, 2008, PEST surface water utilities: Brisbane, Australia, Watermark Numerical Computing and University of Idaho, 141 p., accessed September 28, 2011, at *http://www.pesthomepage.org/getfiles.php?file=swutils.pdf*.

Doherty, John, 2010a, PEST: model-independent parameter estimation user manual (5th ed.): Brisbane, Australia, Watermark Numerical Computing, 336 p.

Doherty, John, 2010b, Addendum to the PEST manual: Brisbane, Australia, Watermark Numerical Computing, 272 p.

Doherty, J., and Johnston, J.M., 2003, Methodologies for calibration and predictive analysis of a watershed model: Journal of the Astronomical Society of Western Australia, v. 39, p. 251–265.

Doherty, J. and Skahill, B.E., 2006, An advanced regularization methodology for use in watershed model calibration: Journal of Hydrology, v. 327, no. 3–4, p. 564–577.

Henricksen, J.A., Heasley, J., Kennen, J.G., and Nieswand, S., 2006, Users' manual for the Hydroecological Integrity Assessment Process software (including the New Jersey Assessment Tools): U.S. Geological Survey Open-File Report 2006–1093, 72 p., accessed at *http://www.fort.usgs.gov/Products/Publications/pub_abstract.asp?PubID=21598*.

Kuczera, G., 1983, Improved parameter inference in catchment models 2. combining different kinds of hydrologic data and testing their compatibility: Water Resources Research, v. 19, no. 5, p. 1163–1172.

Leavesley, G., Lichty, R., Troutman, B., and Saindon, L., 1983, Precipitation-runoff modeling system: user's manual: U.S. Geological Survey Water Resources Investigations Report 83–4238, 207 p.

Legates, D.R., and McCabe, G.J., 1999, Evaluating the use of "goodness-of-fit" measures in hydrologic and hydroclimatic model validation: Water Resources Research, v. 35, no. 1, p. 233–241.

Markstrom, S.L., Niswonger, Richard G., Regan, R. Steven, Prudic, David E., and Barlow, P.M., 2008, GSFLOW—coupled ground-water and surface-water flow model based on the integration of the Precipitation-Runoff Modeling System (PRMS) and the Modular Ground-Water Flow Model (MODFLOW-2005): U.S. Geological Survey Techniques and Methods book 6, chap. D1, 240 p., accessed at *http://pubs.usgs.gov/tm/tm6d1/*.

Muffels, C.T., Schreüder, W.A., Doherty, J.E., Karanovic, M., Tonkin, M.J., Hunt, R.J., and Welter, D.E., 2012, Approaches in highly parameterized inversion—GENIE, a general model-independent TCP/IP run manager: U.S. Geological Survey Techniques and Methods, book 7, section C6, 26 p.

Nash, J.E., and Sutcliffe, J.V., 1970, River flow forecasting through conceptual models part I—a discussion of principles: Journal of Hydrology, v. 10, no. 3, p. 282–290.

Nathan, R.J., and McMahon, T.A., 1990, Evaluation of automated techniques for base flow and recession analyses: Water Resources Research, v. 26, no. 7, p. 1465–1473.

Olden, J.D., and Poff, N.L., 2003, Redundancy and the choice of hydrologic indices for characterizing streamflow regimes: River Research and Applications, v. 19, no. 2, p. 101–121. (Also available at *http://dx.doi.org/10.1002/rra.700*.)

Riggs, H.C., 1972, Low-flow investigations: Techniques of Water-Resources Investigations of the U.S. Geological Survey, book 4, chap. B1, 23 p., accessed at *http://pubs.usgs.gov/twri/twri4b1/*.

Runkel, R.L., Crawford, C.G., and Cohn, T.A., 2004, Load Estimator (LOADEST): a FORTRAN program for estimating constituent loads in streams and rivers: U.S. Geological Survey Techniques and Methods, book 4, chap. A5, 69 p.

Searcy, J.K., 1959, Flow-duration curves, Manual of hydrology, pt. 2, Low-flow techniques: U.S. Geological Survey Water-Supply Paper 1542–A, 32 p. accessed July 30, 2012, at *http://pubs.usgs.gov/wsp/1542a/report.pdf*.

Skahill, B.E., and Doherty, J., 2006, Efficient accommodation of local minima in watershed model calibration: Journal of Hydrology, v. 329, no. 1–2, p. 122–139.

Sloto, R.A., and Crouse, M.Y., 1996, HYSEP: A computer program for streamflow hydrograph separation and analysis: U.S. Geological Survey Water-Resources Investigations Report 96–4040, 46 p.

Welter, D.E., Doherty, J.E., Hunt, R.J., Muffels, C.T., Tonkin, M.J., and Schreüder, W.A., 2012, Approaches in highly parameterized inversion—PEST++, a Parameter ESTimation code optimized for large environmental models: U.S. Geological Survey Techniques and Methods, book 7, section C5, 47 p.

Appendixes 1–3

Appendix 1: File Format Details

Site Sample File (SSF)

The "site sample file" is fundamental to the operation of many of the Surface Water Utilities; it holds time series data gathered at one or *more* sites. The data stored in this file can be of any type.

A site sample file records data, for example, water levels or chemical concentrations, gathered through sampling programs at discrete sample times at a number of specific locations. Each line of a site sample file has four (or possibly five) entries, each of which must be separated from its neighboring entry by one or more white space (including tab) characters. Typically a site sample file holds data extracted from a database. Part of a site sample file is shown in fig. 1–1 below.

```
13500002A   25/09/1991   12:00:00  12.00
13500002A   02/01/1992   12:00:00  11.83
13500002A   24/03/1992   12:00:00  12.81
13500002A   29/06/1992   12:00:00  13.54
13500002A   22/09/1992   12:00:00  13.24
13500002A   17/12/1992   12:00:00  12.84
13500002A   22/03/1993   12:00:00  12.38 x
13500002A   21/06/1993   12:00:00  11.83 x
13500002A   27/09/1993   12:00:00  11.61 x
13500002A   16/12/1993   12:00:00  12.35
13500002A   01/03/1994   12:00:00  11.79
13500002A   22/03/1994   12:00:00  11.89
1351235A    19/02/1959   12:00:00  29.84
1351235A    05/03/1959   12:00:00  30.33
1351235A    20/03/1959   12:00:00  30.76
1351235A    06/04/1959   12:00:00  31.19
1351235A    17/04/1959   12:00:00  31.45
1351235A    01/05/1959   12:00:00  31.65
site_a      15/05/1959   12:00:00  31.65
site_a      29/05/1959   12:00:00  31.65
site_a      12/06/1959   12:00:00  31.65
site_a      26/06/1959   12:00:00  31.46
site_a      10/07/1959   12:00:00  31.34
```

Figure 1–1.   Extract from a site sample file.

The first item on each line of a site sample file is a site identifier. This identifier must be of 18 or fewer characters. When used with programs of the Surface Water Utilities, the site identifier is case insensitive. The second item is the date; depending on the contents of the settings file settings; this must be expressed either in the format dd/mm/yyyy or mm/dd/yyyy (see figure 9 in the previous section of this publication). The observation time (in the format hh mm:ss) and the observation value come third and fourth in each line. An optional fifth item may be present on any line; if present, this item must consist solely of the single character "x" to indicate that the previous data element lacks integrity.

The following rules must be observed when generating a site sample file.

- For any one site, dates and times must be listed in increasing order.

- All entries for the same site must be in juxtaposition; in other words, it is not permitted to list some of the entries for a particular site in one part of a site sample file and the remainder of the entries in another part of the same file, with data pertaining to one or more other sites in between.

- A time entry of 24:00:00 is not permitted; this must be represented as 00:00:00 on the following day.

Statvar File (PRMS Output)

A STATVAR file begins with the number of series N represented in the file. Following that are N lines, each containing a variable name followed by a location identifier, which generally corresponds to a particular hydrologic response unit (HRU), Muskingum stream node, or other reservoir. Taken together, these uniquely identify a series. Following these N lines are the data comprising the time series themselves. The first entry on each such line is the model simulation day. Following that is the year, month, day, hour, minute, and second, respectively, corresponding to series entries on that line followed by the entries themselves. Entries are in the same order as variable name and location identification entries provided in the header to the file.

Modflow/GSFLOW Gage Package Output (*.ggo)

GSFLOW and MODFLOW produce two types of gage file, each with a slightly different header format. One of these lists model-calculated quantities pertaining to surface-water features including lakes and rivers as generated by the MODFLOW streamflow routing (SFR2) process (Niswonger and Prudic, 2005) . The other lists model-calculated quantities computed by the MODFLOW unsaturated zone flow (UZF) process (Niswonger and others, 2006). Both files begin with two headers. The first identifies the gage and, if from the SFR package, the stream segment and reach number. The second header line provides column names. In either case, the file is comprised of multiple columns, each associated with the second header that specifies the information recorded in the column (fig.1–2). One of these columns is elapsed simulation time. Data associated with any of the other columns can be extracted by providing the name of its header.

```
      "GAGE No.  9: K,I,J Coord. =  1,162,259; STREAM SEGMENT =  3; REACH = 26 "
         "DATA:   Time      Stage      Flow      Depth     Width     MP-Flow      Precip.
ET    SFR-Runoff   UZF-Runoff"
          1.0000000E+00  9.0975195E+02  1.1160049E+05  3.5192147E-01  6.1409826E+00
1.0862639E+05  0.0000000E+00  0.0000000E+00  0.0000000E+00  1.7551230E+02
          2.0000000E+00  9.0980646E+02  1.4214981E+05  4.0646315E-01  6.1409826E+00
1.3852216E+05  0.0000000E+00  0.0000000E+00  1.5665702E+03  0.0000000E+00
          3.0000000E+00  9.0992102E+02  2.1525652E+05  5.2100694E-01  6.1409826E+00
2.1028977E+05  0.0000000E+00  0.0000000E+00  4.4122778E+03  0.0000000E+00
```

Figure 1–2. Extract from a MODFLOW gage package output file.

Appendix 2: Python Module Use

The version of TSPROC documented in this report can be compiled as a traditional Fortran program or as a Fortran library with bindings to Python. Many new ways of using and extending TSPROC are possible through the use of the Python module and Fortran library. This is a capability that will likely continue to evolve long after the publication of this report; users should check online documentation for the most current information regarding the capabilities of the Python TSPROC module.

The Fortran wrapping package f2py (Peterson, 2009) has been used to assist in creating entry points to the Fortran code from within Python. The TSPROC code may be used in several different ways from within a Python script. The simplest application invokes TSPROC from within a Python script as opposed to from the command line. A more complex Python script can create TSPROC input blocks in memory, obviating the need for a control file. Lastly, the TSPROC module may be used in such a way that the Python script simply monitors the block currently being processed; this allows the user the option to override or extend current TSPROC functionality. The basic Python module methods are listed in table 2–1.

Table 2–1. Basic Python module methods.

[—, not applicable]

Python script process function name	Arguments	Return values	Description
Work independently of TSPROC control file			
newblock	block_name	—	Deletes any previous in-memory blocks and creates a new block with the name as given by block_name.
addtoblock	block_text	—	Adds the text given in block_text to the in-memory block.
processblock		—	Processes the current block (either in-memory, or in an open TSPROC control file).
Require use of a TSPROC control file			
inittsproc	control_filename, record_filename	—	Sets the name for the control filename and the record filename.
context	context_name	—	Overrides any context setting present in the TSPROC control file.
getnextblockname		[block name]	
getnextblock		[block_keywords], [block_arguments]	Returns the block name and all keyword/argument pairs for the next block; a TSPROC control file must already have been initialized to use this function.
runtsproc		—	Starts a TSPROC run. Assumes all input will be provided by the TSPROC control file, which must be specified by a previous call to inittsproc.
Work with or without an active TSPROC control file			
listtables		[table_names]	Returns a list of all TSPROC tables currently resident in memory.
listseries		[series_names]	Returns a list of all TSPROC series currently resident in memory.
gettable	table_name	[table_fieldnames], [table_values]	Returns a list of table fieldnames, and a list of table values.
getseries	series_name	[series_date-times], [series_values]	Returns a list of series datetimes and a list of series values.
maketable	table_name, table_fieldnames, table_values	—	Creates and adds a G_TABLE to the list of active tables within TSPROC.
makeseries	series_name, series_datetimes, series_values	—	Creates and adds a series to the list of active series within TSPROC.

Figure 2–1 shows a Python script that initializes the TSPROC library and then starts a TSPROC run without further processing from the Python script. This could be of value if Python were used as the scripting language to make up the composite model, rather than as a batch file or shell script.

```
import pytsproc.main loop as tsp

tsp.context("all")
tsp.inittsproc("my tsproc file.ctl", "my tsproc recfile.txt")
tsp.runtsproc()


# other executables could be launched before or after TSPROC
# is invoked
```

Figure 2–1. Example of simply starting TSPROC from witin a Python script.

Figure 2–2 shows a Python script that relies on the traditional TSPROC control file for direction, but compares the current block name to the name of a routine implemented in Python. If the appropriate block name is found, it is executed within the Python script, a new series or table is created and added to the list of series or tables in TSPROC memory, and processing of the control file resumes.

```
import pytsproc.main loop as tsp

tsp.context("all")
tsp.inittsproc("my_tsproc_file.ctl", "my_tsproc_recfile.txt")

blockname = tsp.getnextblockname()

while blockname.strip().upper() != "EOF":

 if(blockname.strip().upper() == "MY_NEW_TSPROC_FUNCTION":
  (keywords, arguments) = tsp.getblock()
# make a call to a new, custom function
# custom function must:
#  - process keywords and arguments
#  - call tsp.getseries or tsp.gettable in order to obtain
#    locally modifiable copies of the series or table
#  - process the series or table accordingly
#  - call tsp.makeseries or tsp.maketable to add the newly
#    created series or table to the series or tables that
#    the TSPROC Fortran module holds in memory
  myNewTSPROCFunction(keywords, arguments)

 else:

# this is a block that presumably can be dealt with by the
# Fortran modules...send control back
  tsp.processblock()
```

Figure 2–2. Example Python script that scans blocks in a TSPROC control file, processing the block for which it has a routine.

Figure 2–3 shows an example of a Python script that does not need a TSPROC control file to direct its processing. Instead, the script creates "in-memory" TSPROC blocks on-the-fly and causes them to be processed by TSPROC as though they originated within a control file.

```
import pytsproc.main_loop as tsp

tsp.context("all")

# assemble a TSPROC block to read a series from an SSF file
tsp.newblock("GET MUL SERIES SSF")
tsp.addtoblock("CONTEXT all")
tsp.addtoblock("FILE 05406500.ssf")
tsp.addtoblock("NEW_SERIES q6500i")

# now that the block has been created, make TSPROC act on it
tsp.processblock()


# assemble a TSPROC block to process a time series using the
# USGS HYSEP routines
tsp.newblock("USGS HYSEP")
tsp.addtoblock("CONTEXT all")
tsp.addtoblock("SERIES q6500i")
tsp.addtoblock("NEW SERIES q6500 hysp")
tsp.addtoblock("HYSEP_TYPE sliding_interval")

# now that the block has been created, make TSPROC act on it
# TSPROC routine
tsp.processblock()
```

Figure 2–3. Example of a Python script that creates TSPROC block on-the-fly (in memory).

From within a Python script, a block may be built up by calling the "newblock" method first, specifying the name of the block to be created, followed by repeated calls to the "addtoblock" method. The block keywords and arguments should be given exactly as specified in the main body of this report. Finally, the "processblock" method should be called; this hands control back to the Fortran TSPROC library, with the appropriate TSPROC routines called as usual.

At first it may not appear useful to be able to piece together the bits of a TSPROC block inside a Python script before having TSPROC act on it. What makes this useful is that it is relatively easy to have Python repeat the same actions for any number of files or time series with almost no additional lines of code. Figure 2–4 gives an example of a Python script that makes a list of all site sample files (SSF) within the current directory, reads and creates a TSPROC time series for each, and then performs a hydrograph separation for each.

The concept employed in figure 2–4 may be used to process an almost unlimited number of files without increasing the size of the script; a dozen or mode more lines of TSPROC keyword and argument pairs might be needed for each additional SSF file processed by means of the traditional control file approach.

```python
import os
import pytsproc.main loop as tsp


# define a function that creates a GET_MUL_SERIES_SSF block
def getmulseriesssf(filename, series):
 tsp.newblock("GET_MUL_SERIES_SSF")
 tsp.addtoblock("CONTEXT all")
 tsp.addtoblock("FILE " + filename)
 tsp.addtoblock("NEW SERIES "+ series)
 tsp.processblock()


# define a function that creates a USGS HYSEP block
def usgshysep(series):
 tsp.newblock("USGS HYSEP")
 tsp.addtoblock("CONTEXT all")
 tsp.addtoblock("SERIES " + series)
 tsp.addtoblock("NEW SERIES " + series + " sep")
 tsp.addtoblock("HYSEP TYPE sliding interval")
 tsp.processblock()


# get a list of all files in current directory
filenames = os.listdir(os.curdir)


# for all *.ssf files in the current directory, read in the
# file and process the series using the USGS Hysep methods
for filename in filenames:
 if 'ssf' in filename:

  # split the filename; name the series after the file prefix
  series = filename.split(".")[0]

  # read the SSF file, create the series
  getmulseriesssf(filename, series)

  # run hydrograph separation on the series
  usgshysep(series)
```

Figure 2–4. Example of Python script that processes all SSF files in the current directory.

Appendix 3: Complete List of Indices Calculated in a HYDROLOGIC_INDICES Block

The following information for the 171 hydrologic indices is adapted from information given in Henrickson and others (2006), which in turn is based on information in Olden and Poff (2003). These definitions, taken from Henrickson and others (2006), differ slightly, but insignificantly, from the indices presented in Olden and Poff (2003).

The alphanumeric code preceding each definition refers to the category of the flow regime (M, magnitude; F, frequency; D, duration; T, timing, RA, rate of change) and type of flow event (A, average; L, low; H, high) that the hydrologic index was developed to describe. Indices are numbered successively within each category. For example, MA1 is the first index describing magnitude of the average flow condition. Table 3–1 lists the 11 general index classes developed originally in Olden and Poff (2003).

Table 3–2 gives the definition of each of the indices available in the TSPROC HYDROLOGIC_INDICES block. Following each definition, in parentheses, are (1) the units of the index, and (2) the type of data, temporal or spatial data, from which the upper and lower percentiles limits (for example, 75/25) are derived. Temporal data are from a multiyear daily flow record from a single stream gage. For example, index MA1—mean for the entire flow record—uses 365 mean daily flow values for each year in the flow record to calculate the mean for the entire flow record. Consequently, 365 values for each year are used to calculate upper and lower percentile limits. However, formulas for 60 of the indices do not produce a range of values from which percentile limits can be calculated. Calculation of MA5 (skewness), for example, involves taking the mean for the entire flow record divided by the median for the entire record and results in a single value; upper and lower percentile limits cannot be calculated. This is not of great concern, because the TSPROC HYDROLOGIC_INDICES block is not set up currently to report these upper and lower percentile limits.

Exceedance and percentile are used in the calculation for a number of indices. Note the difference—a 90-percent exceedance means that 90 percent of the values are equal to or greater than the 90-percent exceedance value, while a 90th percentile means that 10 percent of the values are equal to or greater than the 90th percentile value.

Table 3–1. The 11 general index classes developed originally in Olden and Poff (2003).

Alphanumeric code prefix	Index class
MA#	Magnitude, average-flow event
ML#	Magnitude, low-flow event
MH#	Magnitude, high-flow event
FL#	Frequency, low-flow event
FH#	Frequency, high-flow event
DL#	Duration, low-flow event
DH#	Duration, high-flow event
TA#	Timing, average-flow event
TL#	Timing, low-flow event
TH#	Timing, high- flow event
RA#	Rate of change, average event

Table 3–2.   Definitions of indices available in the TSPROC HYDROLOGIC_INDICES block

[Modified from Henrickson and others (2006); refer to table 3–1 for the alphanumeric codes of the indices; text in the "Units" column describe the units of measurement for each index and the type of data, temporal or spatial, from which the upper and lower percentiles limits are derived]

Index	Definition	Units	Definition altered by the USE_MEDIAN keyword?
Magnitude, average-flow event (MA)			
MA1	Mean of the daily mean flow values for the entire flow record.	cubic feet per second—temporal	
MA2	Median of the daily mean flow values for the entire flow record.	cubic feet per second—temporal	
MA3	Mean (or median) of the coefficients of variation (standard deviation/mean) for each year. Compute the coefficient of variation for each year of daily flows. Compute the mean of the annual coefficients of variation.	percent—temporal	☑
MA4	Standard deviation of the percentiles of the logs of the entire flow record divided by the mean of percentiles of the logs. Compute the log10 of the daily flows for the entire flow record. Compute the 5th, 10th, 15th, 20th, 25th, 30th, 35th, 40th, 45th, 50th, 55th, 60th, 65th, 70th, 75th, 80th, 85th, 90th, and 95th percentiles for the logs of the entire flow record. Percentiles are computed by interpolating between the ordered (ascending) logs of the flow values. Compute the standard deviation and mean for the percentile values. Divide the standard deviation by the mean.	percent—spatial	
MA5	The skewness of the entire flow record is computed as the mean for the entire flow record (MA1) divided by the median (MA2) for the entire flow record.	dimensionless—spatial	
MA6	Range in daily flows is the ratio of the 10-percent to 90-percent exceedance values for the entire flow record. Compute the 5-percent to 95-percent exceedance values for the entire flow record. Exceedance is computed by interpolating between the ordered (descending) flow values. Divide the 10-percent exceedance value by the 90-percent value.	dimensionless—spatial	
MA7	Range in daily flows is computed like MA6, except using the 20 percent and 80 percent exceedance values. Divide the 20 percent exceedance value by the 80 percent value.	dimensionless—spatial	
MA8	Range in daily flows is computed like MA6, except using the 25-percent and 75-percent exceedance values. Divide the 25-percent exceedance value by the 75-percent value.	dimensionless—spatial	
MA9	Spread in daily flows is the ratio of the difference between the 90th and 10th percentile of the logs of the flow data to the log of the median of the entire flow record. Compute the log10 of the daily flows for the entire flow record. Compute the 5th, 10th, 15th, 20th, 25th, 30th, 35th, 40th, 45th, 50th, 55th, 60th, 65th, 70th, 75th, 80th, 85th, 90th, and 95th percentiles for the logs of the entire flow record. Percentiles are computed by interpolating between the ordered (ascending) logs of the flow values. Compute MA9 as (90th −10th) /log10(MA2).	dimensionless—spatial	
MA10	Spread in daily flows is computed like MA9, except using the 20th and 80th percentiles.	dimensionless—spatial	
MA11	Spread in daily flows is computed like MA9, except using the 25th and 75th percentiles.	dimensionless—spatial	

Table 3–2. Definitions of indices available in the TSPROC HYDROLOGIC_INDICES block.—Continued

[Modified from Henrickson and others (2006); refer to table 3–1 for the alphanumeric codes of the indices; text in the "Units" column describe the units of measurement for each index and the type of data, temporal or spatial, from which the upper and lower percentiles limits are derived]

Index	Definition	Units	Definition altered by the USE_MEDIAN keyword?
MA12–MA23	Means (or medians) of monthly flow values. Compute the means for each month over the entire flow record. For example, MA12 is the mean of all January flow values over the entire record (cubic feet per second— temporal).		☑
MA24–MA34	Variability (coefficient of variation) of monthly flow values. Compute the standard deviation for each month in each year over the entire flow record. Divide the standard deviation by the mean for each month. Average (or take median of) these values for each month across all years.	percent— temporal	☑
MA36	Variability across monthly flows. Compute the minimum, maximum, and mean flows for each month in the entire flow record. MA36 is the maximum monthly flow minus the minimum monthly flow divided by the median monthly flow.	dimensionless— spatial	
MA37	Variability across monthly flows. Compute the first (25th percentile) and the third (75th percentile) quartiles (every month in the flow record). MA37 is the third quartile minus the first quartile divided by the median of the monthly means.	dimensionless— spatial	
MA38	Variability across monthly flows. Compute the 10th and 90th percentiles for the monthly means (every month in the flow record). MA38 is the 90th percentile minus the 10th percentile divided by the median of the monthly means.	dimensionless— spatial	
MA39	Variability across monthly flows. Compute the standard deviation for the monthly means. MA39 is the standard deviation times 100 divided by the mean of the monthly means.	percent—spatial	
MA40	Skewness in the monthly flows. MA40 is the mean of the monthly flow means minus the median of the monthly means divided by the median of the monthly means.	dimensionless— spatial	
MA41	Annual runoff. Compute the annual mean daily flows. MA41 is the mean of the annual means divided by the drainage area.	cubic feet per second/ square mile— temporal	
MA42	Variability across annual flows. MA42 is the maximum annual flow minus the minimum annual flow divided by the median annual flow.	dimensionless— spatial	
MA43	Variability across annual flows. Compute the first (25th percentile) and third (75th percentile) quartiles and the 10th and 90th percentiles for the annual means (every year in the flow record). MA43 is the third quartile divided by the first quartile divided by the median of the annual means.	dimensionless— spatial	
MA44	Variability across annual flows. Compute the first (25th percentile) and third (75th percentile) quartiles and the 10th and 90th percentiles for the annual means (every year in the flow record). MA44 is the 90th percentile minus the 10th percentile divided by the median of the annual means.	dimensionless— spatial	
MA45	Skewness in the annual flows. MA45 is the mean of the annual flow means minus the median of the annual means divided by the median of the annual means.	dimensionless — spatial	

Table 3-2.   Definitions of indices available in the TSPROC HYDROLOGIC_INDICES block.—Continued

[Modified from Henrickson and others (2006); refer to table 3–1 for the alphanumeric codes of the indices; text in the "Units" column describe the units of measurement for each index and the type of data, temporal or spatial, from which the upper and lower percentiles limits are derived]

Index	Definition	Units	Definition altered by the USE_MEDIAN keyword?
	Magnitude, low-flow event (ML)		
ML1–ML12	Mean (or median) of minimum flows for each month across all years. Compute the minimums for each month over the entire flow record. For example, ML1 is the mean of the minimums of all January flow values over the entire record.	cubic feet per second—temporal	☑
ML13	Variability (coefficient of variation) across minimum monthly flow values. Compute the mean and standard deviation for the minimum monthly flows over the entire flow record. ML13 is the standard deviation times 100 divided by the mean minimum monthly flow for all years.	percent—spatial	
ML14	Compute the minimum annual flow for each year. ML14 is the mean of the ratios of minimum annual flows to the median flow for each year.	dimensionless—temporal	
ML15	Low-flow index. ML15 is the mean of the ratios of minimum annual flows to the mean flow for each year.	dimensionless—temporal	
ML16	Median of annual minimum flows. ML16 is the median of the ratios of minimum annual flows to the median flow for each year.	dimensionless—temporal	
ML17	Base flow. Compute the minimum of a 7-day moving average flows for each year and divide them by the mean annual flow for that year. ML17 is the mean (or median—Use Preferenceset by using the Preference option) of those ratios.	dimensionless—temporal	☑
ML18	Variability in base flow. Compute the standard deviation for the ratios of 7-day moving average flows to mean annual flows for each year. ML18 is the standard deviation times 100 divided by the mean of the ratios.	percent—spatial	
ML19	Base flow. Compute the ratios of the minimum annual flow to mean annual flow for each year. ML19 is the mean (or median) of these ratios times 100.	dimensionless—temporal	☑
ML20	Base flow. Divide the daily flow record into 5-day blocks. Find the minimum flow for each block. Assign the minimum flow as a base flow for that block if 90 percent of that minimum flow is less than the minimum flows for the blocks on either side. Otherwise, set it to zero. Fill in the zero values using linear interpolation. Compute the total flow for the entire record and the total base flow for the entire record. ML20 is the ratio of total flow to total base flow.	dimensionless—spatial	
ML21	Variability across annual minimum flows. Compute the mean and standard deviation for the annual minimum flows. ML21 is the standard deviation times 100 divided by the mean.	percent—spatial	
ML22	Specific mean annual minimum flow. ML22 is the mean (or median) of the annual minimum flows divided by the drainage area.	cubic feet per second/square mile—temporal	☑

Table 3–2. Definitions of indices available in the TSPROC HYDROLOGIC_INDICES block.—Continued

[Modified from Henrickson and others (2006); refer to table 3–1 for the alphanumeric codes of the indices; text in the "Units" column describe the units of measurement for each index and the type of data, temporal or spatial, from which the upper and lower percentiles limits are derived]

Index	Definition	Units	Definition altered by the USE_MEDIAN keyword?
	Magnitude, high-flow event (MH)		
MH1–MH12	Mean (or median) maximum flows for each month across all years. Compute the maximums for each month over the entire flow record. For example, MH1 is the mean of the maximums of all January flow values over the entire record.	cubic feet per second—temporal	☑
MH13	Variability (coefficient of variation) across maximum monthly flow values. Compute the mean and standard deviation for the maximum monthly flows over the entire flow record. MH13 is the standard deviation times 100 divided by the mean maximum monthly flow for all years.	percent—spatial	
MH14	Median of annual maximum flows. Compute the annual maximum flows from monthly maximum flows. Compute the ratio of annual maximum flow to median annual flow for each year. MH14 is the median of these ratios.	dimensionless—temporal	
MH15	High flow discharge index. Compute the 1-percent exceedance value for the entire data record. MH15 is the 1-percent exceedance value divided by the median flow for the entire record.	dimensionless—spatial	
MH16	High flow discharge index. Compute the 10-percent exceedance value for the entire data record. MH16 is the 10-percent exceedance value divided by the median flow for the entire record.	dimensionless—spatial	
MH17	High flow discharge index. Compute the 25-percent exceedance value for the entire data record. MH17 is the 25-percent exceedance value divided by the median flow for the entire record.	dimensionless—spatial	
MH18	Variability across annual maximum flows. Compute the logs (log10) of the maximum annual flows. Find the standard deviation and mean for these values. MH18 is the standard deviation times 100 divided by the mean.	percent—spatial	
MH19	Skewness in annual maximum flows. Use the equation: $$MH19 = \frac{N^2 \sum q_m^3 - 3N \sum q_m \sum q_m^2 + 2\left(\sum q_m\right)^3}{N(N-1)(N-2)S^3}$$ where: N = number of years $qm = \log_{10}$(annual maximum flows) S = standard deviation of the annual maximum flows.	dimensionless—spatial	
MH20	Specific mean annual maximum flow. MH20 is the mean (or median) of the annual maximum flows divided by the drainage area.	cubic feet per second/square mile—temporal	☑
MH21	High flow volume index. Compute the average volume for flow events above a threshold equal to the median flow for the entire record. MH21 is the average volume divided by the median flow for the entire record.	days—temporal	

Table 3–2. Definitions of indices available in the TSPROC HYDROLOGIC_INDICES block.—Continued

[Modified from Henrickson and others (2006); refer to table 3–1 for the alphanumeric codes of the indices; text in the "Units" column describe the units of measurement for each index and the type of data, temporal or spatial, from which the upper and lower percentiles limits are derived]

Index	Definition	Units	Definition altered by the USE_MEDIAN keyword?
MH22	High flow volume. Compute the average volume for flow events above a threshold equal to three times the median flow for the entire record. MH22 is the average volume divided by the median flow for the entire record.	days—temporal	
MH23	High flow volume. Compute the average volume for flow events above a threshold equal to seven times the median flow for the entire record. MH23 is the average volume divided by the median flow for the entire record.	days—temporal	
MH24	High peak flow. Compute the average peak flow value for flow events above a threshold equal to the median flow for the entire record. MH24 is the average peak flow divided by the median flow for the entire record.	dimensionless—temporal	
MH25	High peak flow. Compute the average peak flow value for flow events above a threshold equal to three times the median flow for the entire record. MH25 is the average peak flow divided by the median flow for the entire record.	dimensionless—temporal	
MH26	High peak flow. Compute the average peak flow value for flow events above a threshold equal to seven times the median flow for the entire record. MH26 is the average peak flow divided by the median flow for the entire record.	dimensionless—temporal	
MH27	High peak flow. Compute the average peak flow value for flow events above a threshold equal to 75th-percentile value for the entire flow record. MH27 is the average peak flow divided by the median flow for the entire record.	dimensionless—temporal	
Frequency, low-flow event (FL)			
FL1	Low flood pulse count. Compute the average number of flow events with flows below a threshold equal to the 25th-percentile value for the entire flow record. FL1 is the average (or median) number of events.	number of events/ year—temporal	☑
FL2	Variability in low pulse count. Compute the standard deviation in the annual pulse counts for FL1. FL2 is 100 times the standard deviation divided by the mean pulse count.	percent—spatial	
FL3	Frequency of low pulse spells. Compute the average number of flow events with flows below a threshold equal to 5 percent of the mean flow value for the entire flow record. FL3 is the average (or median) number of events.	number of events/ year—temporal	☑
Frequency, high-flow event (FH)			
FH1	High flood pulse count. Compute the average number of flow events with flows above a threshold equal to the 75th-percentile value for the entire flow record. FH1 is the average (or median) number of events.	number of events/ year—temporal	☑
FH2	Variability in high pulse count. Compute the standard deviation in the annual pulse counts for FH1. FH2 is 100 times the standard deviation divided by the mean pulse count.	number of events/ year—spatial	
FH3	High flood pulse count. Compute the average number of days per year that the flow is above a threshold equal to three times the median flow for the entire record. FH3 is the mean (or median) of the annual number of days for all years.	number of days/ year—temporal	☑
FH4	High flood pulse count. Compute the average number of days per year that the flow is above a threshold equal to seven times the median flow for the entire record. FH4 is the mean (or median) of the annual number of days for all years.	number of days/ year—temporal	☑

Table 3–2. Definitions of indices available in the TSPROC HYDROLOGIC_INDICES block.—Continued

[Modified from Henrickson and others (2006); refer to table 3–1 for the alphanumeric codes of the indices; text in the "Units" column describe the units of measurement for each index and the type of data, temporal or spatial, from which the upper and lower percentiles limits are derived]

Index	Definition	Units	Definition altered by the USE_MEDIAN keyword?
FH5	Flood frequency. Compute the average number of flow events with flows above a threshold equal to the median flow value for the entire flow record. FH5 is the average (or median) number of events.	number of events/ year—temporal	☑
FH6	Flood frequency. Compute the average number of flow events with flows above a threshold equal to three times the median flow value for the entire flow record. FH6 is the average (or median) number of events.	number of events/ year—temporal	☑
FH7	Flood frequency. Compute the average number of flow events with flows above a threshold equal to seven times the median flow value for the entire flow record. FH6 is the average (or median) number of events.	number of events/ year—temporal	☑
FH8	Flood frequency. Compute the average number of flow events with flows above a threshold equal to 25-percent exceedance value for the entire flow record. FH8 is the average (or median) number of events.	number of events/ year—temporal	☑
FH9	Flood frequency. Compute the average number of flow events with flows above a threshold equal to 75-percent exceedance value for the entire flow record. FH9 is the average (or median) number of events.	number of events/ year—temporal	☑
FH10	Flood frequency. Compute the average number of flow events with flows above a threshold equal to median of the annual minima for the entire flow record. FH10 is the average (or median) number of events (number of events/year—temporal). Note the 1.67-year flood threshold (Poff, 1996) that applies to indices FH11, DH22, DH23, DH24, TA3, and TH3 (below). Compute the log10 of the peak annual flows. Compute the log10 of the daily flows for the peak annual flow days. Calculate the coefficients for a linear regression equation for logs of peak annual flow versus logs of average daily flow for peak days. Using the log peak flow for the 1.67 year recurrence interval (60th percentile) as input to the regression equation, predict the log10 of the average daily flow. The threshold is 10 to the log10 (average daily flow) power.	cubic feet/second	☑
FH11	Flood frequency. Compute the average number of flow events with flows above a threshold equal to flow corresponding to a 1.67-year recurrence interval (Poff, 1996; see index FH10 for computation details). FH11 is the average (or median) number of events.	number of events/ year—temporal	☑
Duration, low-flow event (DL)			
DL1	Annual minimum daily flow. Compute the minimum 1-day average flow for each year. DL1 is the mean (or median) of these values.	cubic feet per second—temporal	☑
DL2	Annual minimum of 3-day moving average flow. Compute the minimum of a 3-day moving average flow for each year. DL2 is the mean (or median) of these values.	cubic feet per second—temporal	☑
DL3	Annual minimum of 7-day moving average flow. Compute the minimum of a 7-day moving average flow for each year. DL3 is the mean (or median) of these values.	cubic feet per second—temporal	☑
DL4	Annual minimum of 30-day moving average flow. Compute the minimum of a 30-day moving average flow for each year. DL4 is the mean (or median) of these values.	cubic feet per second—temporal	☑

Table 3–2. Definitions of indices available in the TSPROC HYDROLOGIC_INDICES block.—Continued

[Modified from Henrickson and others (2006); refer to table 3–1 for the alphanumeric codes of the indices; text in the "Units" column describe the units of measurement for each index and the type of data, temporal or spatial, from which the upper and lower percentiles limits are derived]

Index	Definition	Units	Definition altered by the USE_MEDIAN keyword?
DL5	Annual minimum of 90-day moving average flow. Compute the minimum of a 90-day moving average flow for each year. DL5 is the mean (or median) of these values.	cubic feet per second—temporal	☑
DL6	Variability of annual minimum daily average flow. Compute the standard deviation for the minimum daily average flow. DL6 is 100 times the standard deviation divided by the mean.	percent—spatial	
DL7	Variability of annual minimum of 3-day moving average flow. Compute the standard deviation for the minimum 3-day moving averages. DL7 is 100 times the standard deviation divided by the mean.	percent—spatial	
DL8	Variability of annual minimum of 7-day moving average flow. Compute the standard deviation for the minimum 7-day moving averages. DL8 is 100 times the standard deviation divided by the mean.	percent—spatial	
DL9	Variability of annual minimum of 30-day moving average flow. Compute the standard deviation for the minimum 30-day moving averages. DL9 is 100 times the standard deviation divided by the mean.	percent—spatial	
DL10	Variability of annual minimum of 90-day moving average flow. Compute the standard deviation for the minimum 90-day moving averages. DL10 is 100 times the standard deviation divided by the mean.	percent—spatial	
DL11	Annual minimum daily flow divided by the median for the entire record. Compute the minimum daily flow for the entire record. DL11 is the mean of these values divided by the median for the entire record.	dimensionless—temporal	
DL12	Annual minimum of 7-day moving average flow divided by the median for the entire record. Compute the minimum of a 7-day moving average flow for each year. DL12 is the mean of these values divided by the median for the entire record.	dimensionless—temporal	
DL13	Annual minimum of 30-day moving average flow divided by the median for the entire record. Compute the minimum of a 30-day moving average flow for each year. DL13 is the mean of these values divided by the median for the entire record.	dimensionless—temporal	
DL14	Low exceedance flows. Compute the 75-percent exceedance value for the entire flow record. DL14 is the exceedance value divided by the median for the entire record.	dimensionless—spatial	
DL15	Low exceedance flows. Compute the 90-percent exceedance value for the entire flow record. DL14 is the exceedance value divided by the median for the entire record.	dimensionless—spatial	
DL16	Low flow pulse duration. Compute the average pulse duration for each year for flow events below a threshold equal to the 25th-percentile value for the entire flow record. DL16 is the median of the yearly average durations.	number of days—temporal	
DL17	Variability in low pulse duration. Compute the standard deviation for the yearly average low pulse durations. DL17 is 100 times the standard deviation divided by the mean of the yearly average low pulse durations.	percent—spatial	
DL18	Number of zero-flow days. Count the number of zero-flow days for the entire flow record. DL18 is the mean (or median) annual number of zero flow days.	number of days/year—temporal	☑

Table 3–2. Definitions of indices available in the TSPROC HYDROLOGIC_INDICES block.—Continued

[Modified from Henrickson and others (2006); refer to table 3–1 for the alphanumeric codes of the indices; text in the "Units" column describe the units of measurement for each index and the type of data, temporal or spatial, from which the upper and lower percentiles limits are derived]

Index	Definition	Units	Definition altered by the USE_MEDIAN keyword?
DL19	Variability in the number of zero-flow days. Compute the standard deviation for the annual number of zero-flow days. DL19 is 100 times the standard deviation divided by the mean annual number of zero-flow days.	percent—spatial	
DL20	Number of zero-flow months. While computing the mean monthly flow values, count the number of months when there was no flow over the entire flow record.	percent—spatial	
Duration, high-flow event (DH)			
DH1	Annual maximum daily flow. Compute the maximum of a 1-day moving average flow for each year. DH1 is the mean (or median) of these values.	cubic feet per second—temporal	☑
DH2	Annual maximum of 3-day moving average flows. Compute the maximum of a 3-day moving average flow for each year. DH2 is the mean (or median) of these values.	cubic feet per second—temporal	☑
DH3	Annual maximum of 7-day moving average flows. Compute the maximum of a 7-day moving average flow for each year. DH3 is the mean (or median) of these values.	cubic feet per second—temporal	☑
DH4	Annual maximum of 30-day moving average flows. Compute the maximum of a 30-day moving average flow for each year. DH4 is the mean (or median) of these values.	cubic feet per second—temporal	☑
DH5	Annual maximum of 90-day moving average flows. Compute the maximum of a 90-day moving average flow for each year. DH5 is the mean (or median) of these values.	cubic feet per second—temporal	☑
DH6	Variability of annual maximum daily flows. Compute the standard deviation for the maximum 1-day moving averages. DH6 is 100 times the standard deviation divided by the mean.	percent—spatial	
DH7	Variability of annual maximum of 3-day moving average flows. Compute the standard deviation for the maximum 3-day moving averages. DH7 is 100 times the standard deviation divided by the mean.	percent—spatial	
DH8	Variability of annual maximum of 7-day moving average flows. Compute the standard deviation for the maximum 7-day moving averages. DH8 is 100 times the standard deviation divided by the mean.	percent—spatial	
DH9	Variability of annual maximum of 30-day moving average flows. Compute the standard deviation for the maximum 30-day moving averages. DH9 is 100 times the standard deviation divided by the mean.	percent—spatial	
DH10	Variability of annual maximum of 90-day moving average flows. Compute the standard deviation for the maximum 90-day moving averages. DH10 is 100 times the standard deviation divided by the mean.	percent—spatial	
DH11	Annual maximum of 1-day moving average flows divided by the median for the entire record. Compute the maximum of a 1-day moving average flow for each year. DL11 is the mean of these values divided by the median for the entire record.	dimensionless—temporal	

Table 3–2. Definitions of indices available in the TSPROC HYDROLOGIC_INDICES block.—Continued

[Modified from Henrickson and others (2006); refer to table 3–1 for the alphanumeric codes of the indices; text in the "Units" column describe the units of measurement for each index and the type of data, temporal or spatial, from which the upper and lower percentiles limits are derived]

Index	Definition	Units	Definition altered by the USE_MEDIAN keyword?
DH12	Annual maximum of 7-day moving average flows divided by the median for the entire record. Compute the maximum daily average flow for each year. DL12 is the mean of these values divided by the median for the entire record.	dimensionless—temporal	
DH13	Annual maximum of 30-day moving average flows divided by the median for the entire record. Compute the maximum of a 30-day moving average flow for each year. DL13 is the mean of these values divided by the median for the entire record.	dimensionless—temporal	
DH14	Flood duration. Compute the mean of the mean monthly flow values. Find the 95th percentile for the mean monthly flows. DH14 is the 95th-percentile value divided by the mean of the monthly means.	dimensionless—spatial	
DH15	High flow pulse duration. Compute the average duration for flow events with flows above a threshold equal to the 75th-percentile value for each year in the flow record. DH15 is the median of the yearly average durations.	days/year—temporal	
DH16	Variability in high flow pulse duration. Compute the standard deviation for the yearly average high pulse durations. DH16 is 100 times the standard deviation divided by the mean of the yearly average high pulse durations.	percent—spatial	
DH17	High flow duration. Compute the average duration of flow events with flows above a threshold equal to the median flow value for the entire flow record. DH17 is the average (or median) duration of the events.	days—temporal	☑
DH18	High flow duration. Compute the average duration of flow events with flows above a threshold equal to three times the median flow value for the entire flow record. DH18 is the average (or median) duration of the events.	days—temporal	☑
DH19	High flow duration. Compute the average duration of flow events with flows above a threshold equal to seven times the median flow value for the entire flow record. DH19 is the average (or median) duration of the events.	days—temporal	☑
DH20	High flow duration. Compute the 75th-percentile value for the entire flow record. Compute the average duration of flow events with flows above a threshold equal to the 75th-percentile value for the median annual flows. DH20 is the average (or median) duration of the events.	days—temporal	☑
DH21	High flow duration. Compute the 25th-percentile value for the entire flow record. Compute the average (or median) duration of flow events with flows above a threshold equal to the 25th-percentile value for the entire set of flows. DH21 is the average duration of the events.	days—temporal	☑
DH22	Flood interval. Compute the flood threshold as the flow equivalent for a flood recurrence of 1.67 years. Determine the median number of days between flood events for each year. DH22 is the mean (or median) of the yearly median number of days between flood events.	days—temporal	☑
DH23	Flood duration. Compute the flood threshold as the flow equivalent for a flood recurrence of 1.67 years. Determine the number of days each year that the flow remains above the flood threshold. DH23 is the mean (or median) of the number of flood days for years when floods occur.	days—temporal	☑

Table 3–2. Definitions of indices available in the TSPROC HYDROLOGIC_INDICES block.—Continued

[Modified from Henrickson and others (2006); refer to table 3–1 for the alphanumeric codes of the indices; text in the "Units" column describe the units of measurement for each index and the type of data, temporal or spatial, from which the upper and lower percentiles limits are derived]

Index	Definition	Units	Definition altered by the USE_MEDIAN keyword?
DH24	Flood-free days. Compute the flood threshold as the flow equivalent for a flood recurrence of 1.67 years. Compute the maximum number of days that the flow is below the threshold for each year. DH24 is the mean (or median) of the maximum yearly no-flood days.	days—temporal	☑

Timing, average flow event (TA)

| TA1 | Constancy. Constancy is computed as presented in Colwell (1974). A matrix of values is compiled where the rows are 11 flow categories and the columns are 365 (no February 29th) days of the year. The cell values are the number of times that a flow falls into a category on each day. The categories are: | dimensionless—spatial | |

$$log(flow) < 0.1 \ x \ log(mean.flow)$$
$$0.1 \ x \ log(mean.flow) <= log(flow) < 0.25 \ x \ log(mean.flow)$$
$$0.25 \ x \ log(mean.flow) <= log(flow) < 0.5 \ x \ log(mean.flow)$$
$$0.5 \ x \ log(mean.flow) <= log(flow) < 0.75 \ x \ log(mean.flow)$$
$$0.75 \ x \ log(mean.flow) <= log(flow) < 1.0 \ x \ log(mean.flow)$$
$$1.0x \ log(mean.flow) <= Log(flow) < 1.25 \ x \ log(mean.flow)$$
$$1.25x \ log(mean.flow) <= log(flow) < 1.5 \ x \ log(mean.flow)$$
$$1.5 \ x \ log(mean.flow) <= log(flow) < 1.75 \ x \ log(mean.flow)$$
$$1.75 \ x \ log(mean.flow) <= log(flow) < 2.0 \ x \ log(mean.flow)$$
$$2.0 \ x \ log(mean.flow) <= log(flow) < 2.25 \ x \ log(mean.flow)$$
$$log(flow) >= 2.25 \ x \ log(mean.flow)$$

The row totals, column totals, and grand total are computed. By use of the equations for Shannon information theory parameters, constancy is computed as: 1-(uncertainty with respect to state) log (number of state).

| TA2 | Predictability. Predictability is computed from the same matrix as constancy (see example in Colwell, 1974). It is computed as: 1-(uncertainty with respect to interaction of time and state - uncertainty with respect to time log (number of state). | dimensionless—spatial | |
| TA3 | Seasonal predictability of flooding. Divide years into 2-month periods (that is, Oct–Nov, Dec–Jan, and so forth). Count the number of flood days (flow events with flows > 1.67-year flood) in each period over the entire flow record. TA3 is the maximum number of flood days in any one period divided by the total number of flood days. | dimensionless—temporal | |

Timing, low-flow event (TL)

| TL1 | Julian date of annual minimum. Determine the Julian date of the minimum flow for each water year. Transform the dates to relative values on a circular scale (radians or degrees). Compute the x and y components for each year, and average them across all years. Compute the mean angle as the arc tangent of y-mean divided by x-mean. Transform the resultant angle back to Julian date. | Julian day—spatial | |

Table 3–2. Definitions of indices available in the TSPROC HYDROLOGIC_INDICES block.—Continued

[Modified from Henrickson and others (2006); refer to table 3–1 for the alphanumeric codes of the indices; text in the "Units" column describe the units of measurement for each index and the type of data, temporal or spatial, from which the upper and lower percentiles limits are derived]

Index	Definition	Units	Definition altered by the USE_MEDIAN keyword?
TL2	Variability in Julian date of annual minima. Compute the coefficient of variation for the mean x and y components, and convert to a date (Julian day—spatial). Note the 5-year flood threshold (Poff, 1996) that applies to indices TL3 and TH3. Compute the log10 of the peak annual flows. Compute the log10 of the daily flows for the peak annual flow days. Calculate the coefficients for a linear regression equation for logs of peak annual flow versus logs of average daily flow for peak days. By use of the log peak flow for the 5-year recurrence interval (20th percentile) as input to the regression equation, predict the log10 of the average daily flow. The threshold is 10 to the log10 (average daily flow) power.	cubic feet per second—temporal	
TL3	Seasonal predictability of low flow. Divide years into 2-month periods (that is, Oct–Nov, Dec–Jan, and so forth). Count the number of low flow events (flow events with flows <= 5 year flood threshold) in each period over the entire flow record. TL3 is the maximum number of low flow events in any one period divided by the total number of low flow events.	dimensionless—spatial	
TL4	Seasonal predictability of non-low flow. Compute the number of days that flow is above the 5-year flood threshold as the ratio of number of days to 365 or 366 (leap year) for each year. TL4 is the maximum of the yearly ratios.	dimensionless—spatial	
Timing, high-flow event (TH)			
TH1	Julian date of annual maximum. Determine the Julian date of the maximum flow for each year. Transform the dates to relative values on a circular scale (radians or degrees). Compute the x and y components for each year, and average them across all years. Compute the mean angle as the arc tangent of y-mean divided by x-mean. Transform the resultant angle back to Julian date.	Julian day—spatial	
TH2	Variability in Julian date of annual maxima. Compute the coefficient of variation for the mean x and y components and convert to a date.	Julian days—spatial	
TH3	Seasonal predictability of nonflooding. Computed as the maximum proportion of a 365-day year that the flow is less than the 1.67-year flood threshold and also occurs in all years. Accumulate nonflood days that span all years. TH3 is the maximum length of those flood-free periods divided by 365.	dimensionless—spatial	
Rate of change, average event (RA)			
RA1	Rise rate. Compute the change in flow for days when the change is positive for the entire flow record. RA1 is the mean (or median) of these values.	cubic feet per second/day—temporal	☑
RA2	Variability in rise rate. Compute the standard deviation for the positive flow changes. RA2 is 100 times the standard deviation divided by the mean.	percent—spatial	
RA3	Fall rate. Compute the change in flow for days when the change is negative for the entire flow record. RA3 is the mean (or median) of these values.	cubic feet per second/day—temporal	☑
RA4	Variability in fall rate. Compute the standard deviation for the negative flow changes. RA4 is 100 times the standard deviation divided by the mean.	percent—spatial	

Table 3–2. Definitions of indices available in the TSPROC HYDROLOGIC_INDICES block.—Continued

[Modified from Henrickson and others (2006); refer to table 3–1 for the alphanumeric codes of the indices; text in the "Units" column describe the units of measurement for each index and the type of data, temporal or spatial, from which the upper and lower percentiles limits are derived]

Index	Definition	Units	Definition altered by the USE_MEDIAN keyword?
RA5	Number of day rises. Compute the number of days when the flow is greater than the previous day. RA5 is the number of positive gain days divided by the total number of days in the flow record.	dimensionless—spatial	☑
RA6	Change of flow. Compute the log10 of the flows for the entire flow record. Compute the change in log of flow for days when the change is positive for the entire flow record. RA6 is the median of these values.	cubic feet per second—temporal	☑
RA7	Change of flow. Compute the log10 of the flows for the entire flow record. Compute the change in log of flow for days when the change is negative for the entire flow record. RA7 is the median of these log values.	cubic feet per second/day—temporal	
RA8	Number of reversals. Compute the number of days in each year when the change in flow from one day to the next changes direction. RA8 is the average (or median) of the yearly values.	days—temporal	☑
RA9	Variability in reversals. Compute the standard deviation for the yearly reversal values. RA9 is 100 times the standard deviation divided by the mean.	percent—spatial